Lógos and Máthēma 2

Polish Contemporary Philosophy and Philosophical Humanities

Edited by Jan Hartman

Volume 18

Roman Murawski

Lógos and Máthēma 2

Studies in the Philosophy of Logic
and Mathematics

Bibliographic Information published by the Deutsche Nationalbibliothek
The Deutsche Nationalbibliothek lists this publication in the Deutsche Nationalbibliografie; detailed bibliographic data is available in the internet at http://dnb.d-nb.de.

Library of Congress Cataloging-in-Publication Data
A CIP catalog record for this book has been applied for at the Library of Congress.

This Publication was financially supported by the Adam Mickiewicz University, Poznan.

Printed by CPI books GmbH, Leck

ISSN 2191-1878
ISBN 978-3-631-80714-9 (Print)
E-ISBN 978-3-631-82018-6 (E-Book)
E-ISBN 978-3-631-82019-3 (EPUB)
E-ISBN 978-3-631-82020-9 (MOBI)
DOI 10.3726/b16869

© Peter Lang GmbH
Internationaler Verlag der Wissenschaften
Berlin 2020
All rights reserved.

Peter Lang – Berlin · Bern · Bruxelles · New York ·
Oxford · Warszawa · Wien

All parts of this publication are protected by copyright. Any utilisation outside the strict limits of the copyright law, without the permission of the publisher, is forbidden and liable to prosecution. This applies in particular to reproductions, translations, microfilming, and storage and processing in electronic retrieval systems.

This publication has been peer reviewed.

www.peterlang.com

*Dedicated to
my wife Hania and daughter Zosia*

Foreword

The volume consists of thirteen papers devoted to various problems of the philosophy of logic and mathematics. They can be divided into two groups. The first group contains papers devoted to some general problems of the philosophy of mathematics whereas the second group – papers devoted to the history of logic in Poland and to the work of Polish logicians and mathematicians in the philosophy of mathematics and logic.

The first group is opened by the chapter "On the philosophical meaning of reverse mathematics" in which philosophical consequences of this research project in the foundations of mathematics are discussed. In particular, we are interested in its implications for Hilbert's program. The chapter "On the distinction proof–truth in mathematics" is devoted to some historical, philosophical and logical considerations connected with the distinction between proof and truth in mathematics. The crucial role of Gödel's incompleteness theorems as well as of the undefinability of truth vs. definability of provability and the role of finitary vs. infinitary methods are stressed. The problem of the need of extending the available methods by new rules of inference and new axioms is also considered. The problem of a proof in mathematics is discussed also in the chapter "Some historical, philosophical and methodological remarks on proof in mathematics". It is devoted to historical and philosophical as well as methodological considerations on the role and meaning of proof in mathematics. In particular, the following problems are discussed: the role of informal proofs in mathematical research practice, the concept of formal proof and its role, the problem of the distinction: (formal) provability *vs.* truth and relations between informal and formal proofs. The chapter "The status of Church's Thesis" is devoted to the problem of the status of this famous hypothesis of the theory of computability. The following possibilities are considered and discussed: Church's Thesis as an empirical hypothesis, as an axiom or theorem, as a definition and as an explication. Next two chapters closing the first part of the volume are devoted to particular conceptions in the philosophy of mathematics. The chapter "Between theology and mathematics" is devoted to philosophical and theological as well as mathematical ideas of Nicholas of Cusa (1401–1464). He was a mathematician but first of all a theologian. Connections between theology and philosophy on the one side and mathematics on the other were by him bilateral. In the chapter, it is shown how some theological ideas were used by him to answer the fundamental questions in the philosophy of mathematics. The chapter "Phenomenological ideas in the philosophy of mathematics" is devoted to phenomenological ideas in conceptions of modern philosophy of mathematics. Views of Husserl, Weyl, Becker and Gödel are discussed and analysed. The aim of this chapter is to show the influence of phenomenological ideas on the philosophical conceptions concerning mathematics. We start by indicating the attachment of Edmund Husserl to mathematics and by presenting the main points of his philosophy of mathematics. Next, works of two philosophers who attempted to apply Husserl's phenomenological ideas to the philosophy of mathematics, namely Hermann Weyl and Oskar Becker, are briefly discussed. Lastly, the connections between Husserl's ideas and the philosophy of mathematics of Kurt Gödel are studied.

The second group of chapters is opened by a historical chapter on mathematical logic and the foundations of mathematics in the reborn (after the First World War) Poland. The rise of Warsaw School of Logic and of Polish School of Mathematics are described; and the background of this process, the cultural and scientific, in particular philosophical atmosphere in which those processes took place, is presented. The next chapter in this group "Tarski and his Polish predecessors on truth" is devoted to the description and analysis of Alfred Tarski's views concerning the concept of truth. Conceptions of his Polish predecessors: Twardowski, Łukasiewicz, Zawirski, Czeżowski and Kotarbiński are also discussed. The chapter "Benedykt Bornstein and his philosophy of logic and mathematics" presents philosophical views on logic and mathematics of this rather unknown and almost completely forgotten significant Polish philosopher. Though he was a Ph.D. student of Kazimierz Twardowski, he was not a member of the Lvov-Warsaw Philosophical School – mainly because of his metaphysical views. In some way, he was an individualist; his research did not follow the main trend. However, his views and conceptions were interesting. The following three chapters are devoted to three centres of logic and mathematics in the interwar Poland, namely: Warsaw (Warsaw School of Mathematical Logic), Cracow and Lvov (Lvov School of Mathematics) and to the presentation of philosophical views on logic and mathematics developed and proclaimed there. In particular, views of Tarski, Andrzej Mostowski (Warsaw), Jan Sleszyński, Stanisław Zaremba, Zygmunt Zawirski, Witold Wilkosz and Leon Chwistek (Cracow; in fact Chwistek was active both in Cracow and Lvov) as well as Hugo Steinhaus, Stefan Banach and Eustachy Żyliński (Lvov) are discussed. The volume is closed by a chapter devoted to the description and analysis of philosophical views on logic and mathematics of members of the so-called Cracow Circle. This term is used to describe a group of scholars who tried to apply the methods of modern formal/- mathematical logic to philosophical and theological problems, in particular they attempted to modernize the contemporary Thomism (the trend which was then prevailing) by the logical tools. The group consisted of the Dominican Father Józef (Innocenty) M. Bocheński, Rev. Jan Salamucha, Jan Franciszek Drewnowski as well as the logician Bolesław Sobociński who collaborated with them.

Papers included into this volume (with one exception) were published earlier in journals and collective volumes as separate and independent items. Putting them now together in one volume implies that there appear some unavoidable repetitions. I hope that this circumstance will not be an obstacle for the reader.

I would like to thank all who helped me in the work on this book. First of all, I thank the co-authors who agreed to include into the volume our joint chapters, in particular Professors Thomas Bedürftig and Jan Woleński. I thank also the publishers of particular papers for the permission to reprint them in the present volume. I thank the Faculty of Mathematics and Computer Science of Adam Mickiewicz University in Poznań for the financial support as well as Ms Magdalena Stachowiak for her help in converting some files and Doctor Paweł Mleczko for his advices concerning TEX. Last but not least I thank Mr. Łukasz Gałecki from Peter Lang Verlag for his helpful assistance.

Roman Murawski

Poznań, in June 2019

Contents

On the Philosophical Meaning of Reverse Mathematics 11

On the Distinction Proof–Truth in Mathematics 23

Some Historical, Philosophical and Methodological Remarks on Proof in Mathematics 37

The Status of Church's Thesis (co-author: Jan Woleński) 53

Between Theology and Mathematics. Nicholas of Cusa's Philosophy of Mathematics 69

Phenomenological Ideas in the Philosophy of Mathematics. From Husserl to Gödel (co-author: Thomas Bedürftig) 81

Mathematical Foundations and Logic in Reborn Poland 95

Tarski and his Polish Predecessors on Truth (co-author: Jan Woleński) 107

Benedykt Bornstein's Philosophy of Logic and Mathematics 127

Philosophy of Logic and Mathematics in the Warsaw School of Mathematical Logic 137

The philosophy of Mathematics and Logic in Cracow between the Wars 145

Philosophy of Logic and Mathematics in the Lvov School of Mathematics 159

Cracow Circle and Its Philosophy of Logic and Mathematics 163

Bibliography 181

Source Note 204

Index 205

On the Philosophical Meaning of Reverse Mathematics

The aim of this chapter is to discuss the meaning of some recent results in the foundations of mathematics – more exactly of the so-called reverse mathematics – for the philosophy of mathematics. In particular, we shall be interested in implications of those results for Hilbert's program.

Hilbert's program

One of the reactions on the crisis in the foundations of mathematics on the turn of the 19th century was Hilbert's program. Hilbert's aim was to save the integrity of classical mathematics (dealing with actual infinity) by showing that it is secure.[1] He saw also the supra-mathematical significance of this issue. In 1926 he wrote: "The definite clarification of the nature of the infinite has become necessary, not merely for the special interests of the individual sciences, but for the honor of human understanding itself". Being first of all a mathematician, he "had little patience with philosophy, his own philosophy of mathematics being perhaps best described as naïve optimism – a faith in the mathematician's ability to solve any problem he might set for himself" (cf. Smoryński 1988).

Hilbert's program of clarification and justification of mathematics was Kantian in character (cf. Detlefsen 1993). Following Kant, he claimed that the mathematician's infinity does not correspond to anything in the physical world, that it is "an idea of pure reason" – as Kant used to say. On the other hand, Hilbert wrote in (1926):

> Kant taught – and it is an integral part of his doctrine – that mathematics treats a subject matter which is given independently of logic. Mathematics, therefore can never be grounded solely on logic. Consequently, Frege's and Dedekind's attempts to so ground it were doomed to failure.
> As a further precondition for using logical deduction and carrying out logical operations, something must be given in conception, viz., certain extralogical concrete objects which are intuited as directly experienced prior to all thinking. For logical deduction to be certain, we must be able to see every aspect of these objects, and their properties, differences, sequences, and contiguities must be given, together with the objects themselves, as

---

1 Detlefsen (1990, p. 374) writes that "Hilbert did want to preserve classical mathematics, but this was not for him an end in itself. What he valued in classical mathematics was its efficiency (including its psychological naturalness) as a means of locating the truth of real or finitary mathematics. Hence, any alternative to classical mathematics having the same benefits would presumably have been equally welcome to Hilbert".

something which cannot be reduced to something else and which requires no reduction. This is the basic philosophy which I find necessary not just for mathematics, but for all scientific thinking, understanding and communicating. The subject matter of mathematics is, in accordance with this theory, the concrete symbols themselves whose structure is immediately clear and recognizable.[2]

According to this, Hilbert distinguished between the unproblematic, finitistic part of mathematics and the infinitistic part which needed justification. Finitistic mathematics deals with so-called real sentences, which are completely meaningful because they refer only to given concrete objects. Infinitistic mathematics on the other hand deals with so-called ideal sentences that contain reference to infinite totalities. Hilbert believed that every true finitary proposition had a finitary proof. Infinitistic objects and methods played only an auxiliary role. They enabled us to give easier, shorter and more elegant proofs but every such proof could be replaced by a finitary one. He was convinced that consistency implies existence and that every proof of existence not giving a construction of postulated objects is in fact a presage of such a construction.

Unfortunately, Hilbert did not give a precise definition of finitism – one finds by him only some hints how to understand it. Hence various interpretations are possible (cf., e.g. Detlefsen 1979; Prawitz 1983; Resnik 1974; Smorynski 1988; Tait 1981).

The infinitistic mathematics can be justified only by finitistic methods because only they can give it security (*Sicherheit*). Hilbert's proposal was to base mathematics on finitistic mathematics via proof theory. Its main goal was to show that proofs which use ideal elements in order to prove results in the real part of mathematics always yield correct results. We can distinguish here two aspects: consistency

---

2   „Schon Kant hat gelehrt – und zwar bildet dies einen integrierenden Bestandteil seiner Lehre –, dass die Mathematik über einen unabhängig von aller Logik gesicherten Inhalt verfügt und daher nie und nimmer allein durch Logik begründet werden kann, weshalb auch die Bestrebungen von Frege und Dedekind scheitern mußten. Vielmehr ist als Vorbedingung für die Anwendung logischer Schlüsse und für die Betätigung logischer Operationen schon etwas in der Vorstellung gegeben: gewisse, außer-logische konkrete Objekte, die anschaulich als unmittelbares Erlebnis vor allem Denken da sind. Soll das logische Schließen sicher sein, so müssen sich diese Objekte vollkommen in allen Teilen überblicken lassen und ihre Aufweisung, ihre Unterscheidung, ihr Aufeinanderfolgen oder Nebeneinandergereihtsein ist mit den Objekten zugleich unmittelbar anschaulich gegeben als etwas, das sich nicht noch auf etwas anderes reduzieren läßt oder einer Reduktion bedarf. Dies ist die philosophische Grundeinstellung, die ich für die Mathematik wie überhaupt zu allem wissenschaftlichen Denken, Verstehen und Mitteilen für erforderlich halte. Und insbesondere in der Mathematik sind Gegenstand unserer Betrachtung die konkreten Zeichen selbst, deren Gestalt unserer Einstellung zufolge unmittelbar deutlich und wiedererkennbar ist".

problem and conservation problem[3]. The former consists in showing (by finitistic methods) that the infinitistic mathematics is consistent, and the latter – in showing (again by finitistic methods) that any real sentence which can be proved in the infinitistic part of mathematics can be proved also in the finitistic part, i.e. that infinitistic mathematics is conservative over finitistic mathematics with respect to real sentences and, even more, that there is a finitistic method of translating infinitistic proofs of real sentences into finitistic ones.[4]

Hilbert wanted to solve those problems via his proof theory. His proposal to carry out the program consisted of two steps: 1. to formalize mathematics, i.e., to reconstruct infinitistic mathematics as a big, elaborate formal system and 2. to give a proof of the consistency and conservativeness of mathematics. It should be done by finitistic methods, and it was possible since after formalization one could treat formulas of the system of mathematics as strings of symbols and proofs as strings of formulas (i.e., as strings of strings of symbols).

Incompleteness results

Hilbert and his school had scored some successes in realization of the program of justification of infinite mathematics. In particular, Hilbert's student W. Ackermann showed by finitistic methods the consistency of a fragment of arithmetic of natural numbers (cf. Ackermann 1924–1925, 1940). But soon, something was to happen that undermined Hilbert's program.

We mean here the incompleteness results of Gödel from 1930 which indicated certain cognitive limitations of the deductive methods (cf. Gödel 1931). They showed that one cannot include the whole mathematics in a consistent formalized system based on the first order predicate calculus – what more, one cannot even include in such a system all truth about natural numbers. Even more, no formal theory containing arithmetic of natural numbers can prove its own consistency.

Gödel's results struck Hilbert's program. The question whether they rejected it cannot be answered definitely. The reason is that Hilbert's program was not formulated precisely enough. Hence, various opinions are formulated and defended (cf., e.g. Detlefsen 1979, 1990; Resnik 1974; Smorynski 1977, 1985, 1988). But one thing

---

3   In some of Hilbert's publications (cf., e.g. Hilbert 1926, 1927) both aspects are stressed but usually (cf. Hilbert and Bernays 1934–1939) the one-sided emphasis is put on the consistency problem.
4   Both those aspects are interconnected – as was indicated by G. Kreisel. He showed that if φ is a Π_1^0 sentence and $T \vdash \varphi$ (where T is an infinitistic theory) then $S + Con_T \vdash \varphi$ (where S is a finitistic theory and Con_T is a sentence stating that T is consistent) (cf., e.g. Smoryński 1977). Hence, identifying real sentences with Π_1^0 sentences we see that a solution to the consistency problem yields a solution to the conservation problem.

should be stressed here: the failure of Hilbert's programme for a certain formalized system of arithmetic need not be a failure of Hilbert's programme for elementary number theory in the informal sense. In fact, one cannot exclude the possibility that the latter can be formalized in a system which can be justified on finitistic grounds.[5]

Gödel's true but undecidable (in the formal system of arithmetic) sentence had not mathematical but in fact metamathematical contents (it states: "I am not a theorem"). This diminished the meaning and significance of Gödel's results. There arose a question: Is it possible to indicate examples of true undecidable sentences of mathematical, in particular number–theoretical, contents? Or formal mathematics is complete with respect to sentences which are interesting and reasonable from the mathematical point of view (whatever it means)?

These questions were answered by J. Paris, L. Harrington and L. Kirby who gave examples of true undecidable sentences of combinatorial contents (cf. Paris and Harrington 1977) and of number–theoretical contents (cf. Kirby, Paris 1982). They are examples of true arithmetical sentences without "pure", i.e., arithmetical proofs. More exactly, we got sentences talking about some combinatorial properties of finite sets or properties of sequences of natural numbers but natural proofs of them use infinite sets and, since they cannot be proved in Peano arithmetic PA, each such proof must contain something from beyond the domain of finite objects. Thus they are theorems having "impure" but no "pure" proofs.

Add that those results were used by quasi-empiricists in mathematics who argue that mathematics is not an *a priori* knowledge, it is not absolute and certain, but is rather quasi-empirical, probable and fallible; mathematics is in fact similar to natural sciences. The new undecidable results indicate also that, as E. Post put it, "mathematical proof is in fact a result of creative activity of reason", it is impossible to bound *a priori* the invention of a mathematician.

---

5   Hilbert rejected the opinion that Gödel's results showed the non-executability of his programme. He claimed that they have shown "only that for more advanced consistency proofs one must use the finitistic standpoint in a deeper way than is necessary for the consideration of elementary formalisms" (cf. Hilbert and Bernays 1934–1939, vol. I). Gödel wrote in (1931) that "Theorem XI [i.e. Gödel's second theorem for arithmetic P where P denotes the arithmetic of Peano extended by simple type theory – my remark, R.M.] (and the corresponding results for M and A) [where M is the set theory and A is the analysis – my remark, R.M.] do not contradict Hilbert's formalistic viewpoint. For this viewpoint presupposes only the existence of a consistency proof in which nothing but finitary means of proof is used, and it is conceivable that there exist finitary proofs that cannot be expressed in the formalism of P (or M or A)".

Generalized Hilbert's program

The natural consequence of incompleteness results was the idea of extending the admissible methods and allowing general constructive methods instead of finitistic ones only[6]. This led to the generalized Hilbert's program.

Though the concept of general constructive methods is unprecise, still the idea of broadening of original Hilbert's proof theory has become an accepted paradigm. Investigations were carried out in this direction, and several results were obtained (cf. Feferman 1988; Feferman 1964–1968; Gödel 1958; Kreisel 1958; Schütte 1977; Simpson 1985; Takeuti 1987). We want to note here only that they led to two surprising insights: (a) classical analysis can be formally developed in conservative extensions of elementary number theory and (b) strong impredicative subsystems of analysis can be reduced to constructively meaningful theories, i.e., relative consistency proofs can be given by constructive means for impredicative parts of second order arithmetic.

On the other hand, it should be stressed that all the proposed generalizations of Hilbert's program, are in fact very different from the original Hilbert's proposal. Hilbert's postulate was the validation and justification of classical mathematics by a reduction to finitistic mathematics. The latter was important here for philosophical and, say, ideological reasons: finitistic objects and reasonings have clear physical meaning and are indispensable for all scientific thought. None of the proposed generalizations can be viewed as finitistic (whatever it means). Hence they have another value and meaning from the methodological and generally philosophical point of view. They are not contributing directly to Hilbert's program, but on the other hand they are in our opinion compatible with Hilbert's reductionist philosophy.

Reverse mathematics vs. Hilbert's program

Another consequence of incompleteness results (besides those described above) is so-called relativized Hilbert's program. If the entire infinitistic mathematics cannot be reduced to and justified by finitistic mathematics then one can ask for which part of it is that possible? In other words: How much of infinitistic mathematics can

---

[6] It seems that Bernays was among the first who recognized this need. He wrote: "It thus became apparent that the 'finite Standpunkt' is not the only alternative to classical ways of reasoning and is not necessary implied by the idea of proof theory. An enlarging of the methods of proof theory was therefore suggested: instead of a restriction to finitist methods of reasoning, it was required only that the arguments be of a constructive character, allowing us to deal with more general forms of inference" (cf. Bernays 1967, p. 502).

be developed within formal systems which are conservative over finitistic mathematics with respect to real sentences? This constitutes the relativized version of Hilbert's program. In what follows, we would like to show how the so-called reverse mathematics of Friedman and Simpson contributes to it, providing us with a partial realization of Hilbert's original program.

To be able to consider the issue we should first specify what is exactly meant by finitistic mathematics and by real sentences. We shall follow Tait (1981) where it is claimed that Hilbert's finitism is captured by the formal system of primitive recursive arithmetic[7] (PRA, also called Skolem arithmetic). By real sentences we shall mean Π_1^0 sentences of the language of Peano arithmetic PA.

It turns out that one can formalize classical mathematics not only in set theory, but most of its parts (such as geometry, number theory, analysis, differential equations, complex analysis, etc.) can be formalized in a weaker system called second order arithmetic A_2^- (denoted also sometimes by Z_2).[8] This is a system formalized in a language $L(A_2^-)$ with two sorts of variables: number variables $x, y, z, \ldots$ and set variables $X, Y, Z, \ldots$ Its nonlogical constants are those of Peano arithmetic PA, i.e., $0, S, +, \cdot$ and the membership relation $\in$. Nonlogical axioms of A_2^- are the following:

1. axioms of PA without the axiom scheme of induction,
2. (extensionality) $\forall x (x \in X \equiv x \in Y) \longrightarrow (X = Y)$,
3. (induction axiom) $0 \in X \ \& \ \forall x (x \in X \longrightarrow S(x) \in X) \longrightarrow \forall x (x \in X)$,
4. (axiom scheme of comprehension) $\exists X \forall x (x \in X \equiv \varphi(x, \ldots))$, where $\varphi(x, \ldots)$ is any formula of the language $L(A_2^-)$ (possibly with free number- or set-variables) in which X does not occur free.

Possible models of A_2^- are structures of the form $(\mathcal{X}, \mathcal{M}, \in)$ where $\mathcal{M}$ is a model of PA and $\mathcal{X}$ is a family of subsets of the universe of the model $\mathcal{M}$. (More information on A_2^- and its models can be found in Apt, Marek (1974), and Murawski (1976–1977, 1984a).)

Second order arithmetic is a nice system because one avoids here troubles connected with set theory (in which mathematics is usually formalized), and on the other hand it is strong enough to prove many important theorems of classical mathematics. There is only a problem of impredicativity[9] of A_2^- (connected with the axiom scheme of comprehension, where φ may be any formula of the language of

---

7   For the description of PRA see, e.g. Smoryński (1977).
8   This was first observed by Hilbert and Bernays. Weyl (1918) had shown that a substantial part of ordinary mathematics can be developed within a certain "predicative" subsystem of Z_2 (allowing ω-iterated arithmetical definability).
9   Recall that H. Poincaré saw the source of antinomies in mathematics just in impredicativity and therefore demanded a restriction to predicative methods only.

A_2^-, hence in particular φ may be of the form $\forall Y \psi(Y, x, \ldots)$). But it turns out that in many cases certain fragments of A_2^- suffice, i.e., only particular special forms of the comprehension axiom are needed.

At the Congress of Mathematicians in Vancouver in 1974 H.Friedman formulated a program of foundations of mathematics called today "reverse mathematics" (cf. Friedman 1975). Its aim is to study the role of set existence axioms, i.e., comprehension axioms in ordinary mathematics. The main problem is: Given a specific theorem τ of ordinary mathematics, which set existence axioms are needed in order to prove τ? This research program turned out to be very fruitful and led to many interesting results.[10]

The procedure used in the reverse mathematics (it reveals the inspiration for its name) is the following: Assume we know that a given theorem τ can be proved in a particular fragment $S(\tau)$ of A_2^-. Is $S(\tau)$ the weakest fragment with this property? To answer this question positively, one shows that the principal set existence axiom of $S(\tau)$ is equivalent to τ, the equivalence being provable in some weaker system in which τ itself is not provable. Thus reverse mathematics is, from the point of view of the philosophy of mathematics, an example of a reductionist program with a firm mathematical basis.

Some specific systems – fragments of A_2^- – arose in this context; the most important are: RCA_0, WKL_0, ACA_0, ATR_0 and $\Pi_1^1 - CA_0$. We shall describe only the first three of them.

The system RCA_0 is a theory in the language of A_2^- based on the following axioms: (i) PA^- (i.e., axioms of Peano arithmetic PA without the axiom scheme of induction), (ii) scheme of induction for Σ_1^0 formulas,[11] i.e.:

$$\varphi(o) \,\&\, \forall x [\varphi(x) \longrightarrow \varphi(Sx)] \longrightarrow \forall x \varphi(x),$$

where φ is a Σ_1^0 formula, (iii) (recursive comprehension axiom)

$$\forall x [\varphi(x) \equiv \psi(x)] \longrightarrow \exists X \forall x [x \in X \equiv \varphi(x)],$$

where φ is Σ_1^0 and ψ is Π_1^0. [Axiom (iii) explains the name RCA_0 of the theory.] It can be shown that $(Rec, \mathcal{N}_0, \in)$, where $\mathcal{N}_0$ is the standard model of Peano arithmetic PA and Rec is the family of all recursive sets, is a model of RCA_0.

The theory WKL_0 consists of RCA_0 plus a further axiom known as weak König's lemma (therefore the name WKL_0) which states that every infinite binary tree has an infinite path (this can be formulated in the language of A_2^- using coding). It is

---

10 Drake claims even that the implications of the results of reverse mathematics "make much of what was written in the past on the philosophy of mathematics, obsolete" (cf. Drake 1989).
11 For the definition of and basic information on the arithmetical and analytical hierarchies of formulas and relations see, e.g. Shoenfield (1967).

stronger than RCA_0 what follows, e.g. from the fact that $(Rec, \mathcal{N}_0, \in)$ is not a model of WKL_0 (this is a consequence of Gödel's theorem on essential undecidability of Peano arithmetic).

The theory ACA_0 is PA^- plus induction axiom plus arithmetical comprehension, i.e., comprehension scheme for any arithmetical formula (possibly containing set parameters). This theory is not weaker than WKL_0 because it proves weak König's lemma. It is in fact stronger than WKL_0 what follows from the fact that there are models of WKL_0 consisting of sets definable in $\mathcal{N}_0$ by formulas of a given class (e.g. the family of Δ_2^0 definable sets), whereas for any model $(\mathcal{X}, \mathcal{N}_0, \in)$ of ACA_0 the family $\mathcal{X}$ must be closed with respect to arithmetical definability.

The specified subsystems of A_2^- are appropriate for particular parts of classical mathematics. In RCA_0 one can construct reals, define notions of the convergence of a sequence, of a continuous function, of Riemann's integrability, etc. and prove positive results of recursive analysis and recursive algebra. For example one can prove in RCA_0 that every countable field has an algebraic closure, that every countable ordered field has an extension to a real closed field as well as the intermediate value theorem for continuous functions (cf. Simpson 1998).

The theory WKL_0 turns out to be a quite strong theory, in particular one can prove in it the following theorems:

- the Heine–Borel covering theorem (every covering of $[0,1]$ by a countable sequence of open intervals has a finite subcovering) (cf. Friedman 1976),
- every continuous function on $[0,1]$ is uniformly continuous (cf. Simpson 1998),
- every continuous function on $[0,1]$ is bounded (cf. Simpson 1998),
- every continuous function on $[0,1]$ has a supremum (cf. Simpson 1998),
- every uniformly continuous function on $[0,1]$, which has a supremum, attains it (cf. Simpson 1998),
- every continuous function on $[0,1]$ attains a maximum value (cf. Simpson 1998),
- the Hahn–Banach theorem (cf. Brown, Simpson 1986 and Brown 1987),
- the Cauchy–Peano theorem on the existence of solutions of ordinary differential equations (cf. Simpson 1984),
- every countable commutative ring has a prime ideal (cf. Friedman, Simpson, Smith 1983),
- every countable formally real field can be ordered (cf. Friedman, Simpson, Smith 1983),
- every countable formally real field has a real closure (cf. Friedman, Simpson, Smith 1983),
- Gödel's completeness theorem for the predicate calculus (cf. Friedman 1976 and Simpson 1998).

Even more: if S is one of the above stated theorems then $RCA_0 + S$ is equivalent to WKL_0.

To indicate the strength of ACA_0 let us mention that the following theorems can be proved in it:

- the Bolzano–Weierstrass theorem (every bounded sequence of real numbers has a convergent subsequence) (cf. Friedman 1976),
- every Cauchy sequence of reals is convergent (cf. Simpson 1998),
- every bounded sequence of reals has a supremum (cf. Friedman 1976),
- every bounded increasing sequence of real numbers is convergent (cf. Friedman 1976),
- the Arzela–Ascoli lemma (any bounded equicontinuous sequence of functions on $[0,1]$ has a uniformly convergent subsequence) (cf. Simpson 1998),
- every countable vector space has a basis (cf. Friedman, Simpson, Smith 1983),
- every countable commutative ring has a maximal ideal (cf. Friedman, Simpson, Smith 1983) and
- every countable Abelian group has a unique divisible closure (cf. Friedman, Simpson, Smith 1983).

And again, if S is any of those theorems then $RCA_0 + S$ is equivalent to ACA_0.

What is the meaning of those results? First of all they indicate how much of A_2^- we need in fact to prove various particular theorems of classical mathematics. And it is interesting from the philosophical point of view that certain particular fragments of A_2^- (described above) turn out to be adequate with respect to ordinary mathematical practice. But note that proofs (in the formalized subsystems of the second order arithmetic) of uniqueness are usually more difficult and more complicated than proofs of the existence (in mathematical practice the former are usually simple consequences of the latter). There is also no direct connection between the complexity of a classical proof of a theorem and the level in the hierarchy of subsystems of A_2^- in which a formalized version of it can be proved (as an example can serve here the theorem that every Abelian group has a torsion subgroup which is trivial in classical algebra but RCA_0 + this theorem is equivalent to ACA_0 hence is not a theorem of, say, WKL_0).

In mathematical practice, we encounter often the following situation: Assume that certain theorem τ can be proved in set theory. The natural question is now: Can τ be proved in an elementary way? Observe that if τ can be classified in the hierarchy of subsystems of A_2^- at a level higher than RCA_0 then the answer is negative, i.e., an abstract, infinitistic proof of τ is indispensable and necessary.

Results of reverse mathematics have also interesting mathematical, and not only logical, applications. As an example can serve here Cauchy–Peano theorem on the existence of solutions of ordinary differential equations. Since it is equivalent to

WKL$_0$ over RCA$_0$ and since the structure $(\mathcal{N}_0, \text{Rec}, \in)$ is not a model of WKL$_0$, we get that there exists a differential equation $y' = f(x, y)$, where f is a recursive continuous function, such that it has no recursive solution.

Observe that not every mathematical theorem can be classified in Friedman's hierarchy of subsystems of A_2^-. As an example one can give here Hilbert's basis theorem (cf. Simpson 1988) which states that for any field K and any natural number n all ideals in the ring $K[x_1, \ldots, x_n]$ can be finitely generated. This theorem is provable in ACA$_0$ but RCA$_0$ + "Hilbert's basis theorem" is not equivalent to any of the considered systems.[12]

Add also that there are sentences which are unprovable in the full system A_2^- of second order arithmetic but can be proved in Zermelo-Fraenkel set theory (cf. Friedman 1981).

To indicate the connections of reverse mathematics with Hilbert's program we must recall some results. In the early 1980s, L.Harrington and Z.Ratajczyk proved a theorem on conservativeness of WKL$_0$ (none of them published it; the proof can be found in Simpson 1998).

Theorem 1 *If $(\mathcal{X}, \mathcal{M}, \in)$ is a countable model of RCA$_0$ and $A \in \mathcal{X}$ then there exists a family $\mathcal{Y} \subseteq \mathcal{P}(M)$ such that $A \in \mathcal{Y}$ and $(\mathcal{Y}, \mathcal{M}, \in)$ is a model of* WKL$_0$.

Corollary 2 *The theory* WKL$_0$ *is conservative over* RCA$_0$ *with respect to* Π_1^1 *sentences, i.e., every* Π_1^1 *sentence provable in* WKL$_0$ *can be proved in* RCA$_0$.

What more, Friedman proved in 1977 (this result was not published; it can be found also in Kirby, Paris 1977) that WKL$_0$ is a conservative extension of Skolem arithmetic PRA with respect to Π_2^0 sentences. His proof used model-theoretical methods. W. Sieg improved this result, giving an alternative proof which uses Gentzen-style methods and exhibiting a primitive recursive proof tranformation. Hence the reducibility of WKL$_0$ to PRA is itself provable in PRA.

Combining those results together with the fact that WKL$_0$ is a fairly strong theory (what was indicated above) one comes to the conclusion that a large and significant part of classical mathematics is finitistically reducible. This means in fact that Hilbert's program can be partially realized!

Add that all this has also some "practical" consequences. First observe that the class of Π_2^0 sentences is rather broad – many theorems of number theory can be

---

12 Denote by $WO(\alpha)$ the sentence stating that the ordinal α is a well-ordered set. One can prove in RCA$_0$ that $WO(\omega^\omega)$ is equivalent to Hilbert's basis theorem. But the sentence $WO(\omega^\omega)$ is incomparable with WKL$_0$. On the other hand, for any given natural number n one can prove in RCA$_0$ that $WO(\omega^n)$ is equivalent to Hilbert's basis theorem for n. What more, $WO(\omega^n)$ is provable in RCA$_0$. So we have here certain analogy with the ω-incompleteness of Peano arithmetic (cf. Gödel 1931, see also Mendelson 1970).

formulated as sentences belonging to that class. Since one can formalize within WKL$_0$ the technique of contour integration, hence any Π_2^0 number–theoretic theorem which is provable with the help of it can also be proved "elementarily", i.e., within PRA and, even more, one can effectively (at least theoretically) find such an "elementary" proof. To give one more example, consider Artin's theorem (being a solution to Hilbert's 17th problem[13]). It can be written as a Π_2^0 sentence. Since all results of the theory of real closed fields needed in the proof of Artin's theorem are provable in WKL$_0$, it follows by Friedman's and Sieg's theorems that Artin's theorem can be proved in PRA, i.e., in an elementary way.

It seems that Hilbert would be satisfied by such results!

---

13   Hilbert asked in his 17th problem "whether every definite form [of any number of variables with real coefficients – my remark, R.M.] may not be expressed as a quotient of sums of squares of forms" (cf. Hilbert 1901, see also Browder 1976). Recall that a form is called "definite" if it becomes negative for no real values of the variables. In 1926, Artin answered this question positively.

On the Distinction Proof–Truth in Mathematics

Concepts of proof and truth are (even in mathematics) ambiguous. It is commonly accepted that proof is the ultimate warrant for a mathematical proposition, that proof is a source of truth in mathematics. One can say that a proposition A is true if it holds in a considered structure or if we can prove it. But what is a proof? And what is truth?

The axiomatic method was considered (since Plato, Aristotle and Euclid) to be the best method to justify and to organize mathematical knowledge. The first mature and most representative example of its usage in mathematics were *Elements* of Euclid. They established a pattern of a scientific theory and a paradigm in mathematics. Since Euclid till the end of the 19th century, mathematics was developed as an axiomatic (in fact rather a quasi-axiomatic) theory based on axioms and postulates. Proofs of theorems contained several gaps – in fact the lists of axioms and postulates were not complete, one freely used in proofs various "obvious" truths or referred to the intuition. Proofs were informal and intuitive, they were rather demonstrations; and the very concept of a proof was of a psychological (and not of a logical) nature. Note that almost no attention was paid to the precization and specification of the language of theories – in fact the language of theories was simply the unprecise colloquial language. One should also note here that in fact till the end of the 19th century, mathematicians were convinced that axioms and postulates should be true statements. It seems to be connected with Aristotle's view that a proposition is demonstrated (proved to be true) by showing that it is a logical consequence of propositions already known to be true. Demonstration was conceived here of as a deduction whose premises are known to be true, and a deduction was conceived of as a chaining of immediate inferences.

Basic concepts underlying the Euclidean paradigm have been clarified on the turn of the 19th century. In particular, the intuitive (and rather psychological in nature) concept of an informal proof (demonstration) was replaced by a precise notion of a formal proof and of a consequence. This was the result of the development of mathematical logic and of a crisis of the foundations of mathematics on the turn of the 19th century which stimulated foundational investigations.

One of the directions of those foundational investigations was the program of David Hilbert and his *Beweistheorie*. Note at the very beginning that "this program was never intended as a comprehensive philosophy of mathematics; its purpose was instead to legitimate the entire corpus of mathematical knowledge" (cf. Rowe 1989, p. 200). Note also that Hilbert's views were changing over the years, but always took a formalist direction.

Hilbert sought to justify mathematical theories by means of formal systems, i.e., using the axiomatic method. He viewed the latter as holding the key to a systematic organization of any sufficiently developed subject. In "Axiomatisches Denken" (1918, p. 405) Hilbert wrote:

> When we put together the facts of a given more or less comprehensive field of our knowledge, then we notice soon that those facts can be ordered. This ordering is always introduced with the help of a certain network of concepts (*Fachwerk von Begriffen*) in such a way that to every object of the given field corresponds a concept of this network and to every fact within this field corresponds a logical relation between concepts. The network of concepts is nothing else than the *theory* of the field of knowledge.[14]

By Hilbert the formal frames were contentually motivated. First-order theories were viewed by him together with suitable non-empty domains, *Bereiche*, which indicated the range of the individual variables of the theory and the interpretations of the nonlogical vocabulary. Hilbert, as a mathematician, was not interested in establishing precisely the ontological status of mathematical objects. Moreover, one can say that his program was calling on people to turn their mathematical and philosophical attention away from the problem of the object of mathematical theories and turn it toward a critical examination of the methods and assertions of theories. On the other hand, he was aware that once a formal theory has been constructed, it can admit various interpretations. Recall here his famous sentence from a letter to G. Frege of 29th December 1899 (cf. Frege 1976, p. 67):

> Yes, it is evident that one can treat any such theory only as a network or schema of concepts besides their necessary interrelations, and to think of basic elements as being any objects. If I think of my points as being any system of objects, for example the system: love, law, chimney-sweep [...], and I treat my axioms as [expressing] interconnections between those objects, then my theorems, e.g. the theorem of Pythagoras, hold also for those things. In other words: any such theory can always be applied to infinitely many systems of basic elements.[15]

---

14 „Wenn wir die Tatsachen eines bestimmten mehr oder minder umfassenden Wissensgebiete zusammenstellen, so bemerken wir bald, daß diese Tatsachen einer Ordnung fähig sind. Diese Ordnung erfolgt jedesmal mit Hilfe eines gewissen *Fachwerkes von Begriffen* in der Weise, daß dem einzelnen Gegenstande des Wissensgebietes ein Begriff dieses Fachwerkes und jeder Tatsache innerhalb des Wissensgebietes eine logische Beziehung zwischen den Begriffen entspricht. Das Fachwerk der Begriffe ist nicht Anderes als die *Theorie* des Wissensgebietes".

15 „Ja, es ist doch selbstverständlich eine jede Theorie nur ein Fachwerk oder Schema von Begriffen nebst ihren nothwendigen Beziehungen zu einander, und die Grundelemente können in beliebiger Weise gedacht werden. Wenn ich unter meinen Punkten irgendwelche Systeme von Dingen, z.B. das System: Liebe, Gesetz, Schornsteinfeger [...] denke und dann nur meine sämtlichen Axiome als Beziehungen zwischen diesen Dingen annehme, so gelten meine Sätze, z.B. der Pythagoras auch von diesen Dingen.

The essence of the axiomatic study of mathematical truths consisted for him in the clarification of the position of a given theorem (truth) within the given axiomatic system and of the logical interconnections between propositions.

Hilbert sought to secure the validity of mathematical knowledge by syntactical considerations without appeal to semantic ones. The basis of his approach was the distinction between the unproblematic "finitistic" part of mathematics and the "infinitistic" part that needed justification. As is well known, Hilbert proposed to base mathematics on finitistic mathematics via proof theory (*Beweistheorie*). The latter was planned as a new mathematical discipline in which mathematical proofs are studied by mathematical methods. Its main goal was to show that proofs which use ideal elements (in particular actual infinity) in order to prove results in the real part of mathematics always yield correct results. One can distinguish here two aspects: consistency problem and conservation problem. The consistency problem consists in showing (by finitistic methods, of course) that the infinitistic mathematics is consistent; the conservation problem consists in showing by finitistic methods that any real sentence which can be proved in the infinitistic part of mathematics can be proved also in the finitistic part. One should stress here the emphasis on consistency (instead of correctness).

To realize this program, one should formalize mathematical theories (even the whole of mathematics) and then study them as systems of symbols governed by specified and fixed combinatorial rules.

The formal, axiomatic system should satisfy three conditions: it should be complete, consistent and based on independent axioms. The consistency of a given system was the criterion for mathematical truth and for the very existence of mathematical objects. It was also presumed that any consistent theory would be categorical, i.e., would (up to isomorphism) characterize a unique domain of objects. This demand was connected with the completeness.

The meaning and understanding of completeness by Hilbert plays a crucial role from the point of view of our subject. Note at the beginning that in the *Grundlagen der Geometrie* completeness was postulated as one of the axioms (the axiom was not present in the first edition, but was included first in the French translation and then in the second edition of 1903). In fact the axiom V(2) stated that it is not possible to extend the system of points, lines and planes by adding new entities so that the other axioms are still satisfied. In Hilbert's lecture at the Congress at Heidelberg in 1904 (cf. 1905b), one finds such an axiom system for the real numbers. Later, there appears completeness as a property of a system. In lectures "Logische Principien des mathematischen Denkens" (1905a, p. 13) Hilbert explains the demand

---

Mit anderen Worten: eine jede Theorie kann stets auf unendliche viele Systeme von Grundelementen angewandt werden".

of the completeness as the demand that the axioms suffice to prove all "facts" of the theory in question. He says: "We will have to demand that all other facts (*Thatsachen*) of the given field are consequences of the axioms". On the other hand, one can say that Hilbert's early conviction as to the solvability of every mathematical problem – expressed, e.g. in his 1900 Paris lecture (cf. Hilbert 1901) and repeated in his opening address "Naturerkennen und Logik" (cf. Hilbert 1930b) before the Society of German Scientists and Physicians in Königsberg in September 1930 – can be treated as informal reflection of a belief in completeness.

In his 1900 Paris lecture, Hilbert spoke about completeness in the following words (see the second problem): "When we are engaged in investigating the foundations of a science, we must set up a system of axioms which contains an exact and complete description of the relations subsisting between the elementary ideas of that science".

One can take the "exact and complete description" to be complete enough to decide the truth or falsity of every statement. Semantically such completeness follows from categoricity, i.e., from the fact that any two models of a given axiomatic system are isomorphic; syntactically it means that every sentence or its negation is derivable from the given axioms. Hilbert's own axiomatizations were complete in the sense of being categorical. But notice that they were not first-order, indeed his axiomatization of geometry from *Grundlagen* as well as his axiomatization of arithmetic published in 1900 were second-order.

The demand discussed here would imply that a complete (in this sense) system of axioms is possible only for sufficiently advanced theories. On the other hand, Hilbert called for complete systems of axioms also for theories being developed. One should also add here that Hilbert admitted the possibility that a mathematical problem may have a negative solution, i.e., that one can show the impossibility of a positive solution on the basis of a considered axiom system (cf. Hilbert 1901).

In Hilbert's lectures from 1917–1918 (cf. Hilbert 1917–1918), one finds completeness in the sense of maximal consistency, i.e., a system is complete if and only if for any non-derivable sentence, if it is added to the system then the system becomes inconsistent. In his lecture at the International Congress of Mathematicians in Bologna in 1928, Hilbert stated two problems of completeness: one for the first-order predicate calculus (completeness with respect to validity in all interpretations, hence the semantic completeness) and the second for a system of elementary number theory (formal completeness, in the sense of maximal consistency, i.e., Post-completeness, hence the syntactical completeness) (cf. Hilbert 1930a).

Hilbert's emphasis on the finitary and syntactical methods together with the demand of (and belief in) the completeness of formal systems seem to be the source and reason of the fact that, as Gödel put it (cf. Wang 1974, p. 9), "[...] formalists considered formal demonstrability to be an analysis of the concept of mathematical truth and, therefore were of course not in a position to distinguish the two". Indeed, the informal concept of truth was not commonly accepted as a definite mathematical notion at that time. As Gödel wrote in a crossed-out passage of a draft of his reply to a letter of the student Yossef Balas: "[...] a concept of objective mathematical truth as opposed to demonstrability was viewed with greatest suspicion and widely rejected as meaningless" (cf. Wang 1987, pp. 84–85). Therefore, Hilbert preferred to deal in his metamathematics solely with the forms of the formulas, using only finitary reasonings which were considered to be safe – contrary to semantical reasonings which were non-finitary and consequently not safe. Non-finitary reasonings in mathematics were considered to be meaningful only to the extent to which they could be interpreted or justified in terms of finitary metamathematics.[16]

On the other hand, there was no clear distinction between syntax and semantics at that time. Recall, e.g., that as indicated earlier, the axiom systems came by Hilbert often with a built-in interpretation. Add also that the very notions necessary to formulate properly the difference syntax-semantics were not available to Hilbert.

The problem of the completeness of the first-order logic, i.e., the fourth problem of Hilbert's Bologna lecture, was also posed as a question in the book by Hilbert and Ackermann *Gnmdzüge der theoretischen Logik* (1928). It was solved by Kurt Gödel in his doctoral dissertation (1929, cf. also 1930) where he showed that the first-order logic is complete, i.e., every true statement can be derived from the axioms. Moreover he proved that, in the first-order logic, every consistent axiom system has a model. More exactly, Gödel wrote that by completeness he meant that "every valid formula expressible in the restricted functional calculus [...] can be derived from the axioms by means of a finite sequence of formal inferences". And added that this is equivalent to the assertion that "Every consistent axiom system [formalized within that restricted calculus] [...] has a realization" and to the statement that "Every logical expression is either satisfiable or refutable" (this is the form in which he actually proved the result). The importance of this result is, according to Gödel, that it justifies the "usual method of proving consistency". One should notice here that the notion of truth in a structure, central to the very definition of satisfiability or validity, was nowhere analysed in either Gödel's dissertation or his published revision of it. There was in fact a long tradition of using the informal notion of satisfiability (compare the work of Löwenheim, Skolem and others).

Some months later, in 1930, Gödel solved three other problems posed by Hilbert in Bologna by showing that arithmetic of natural numbers and all richer

---

16 Cf. Gödel's letter to Hao Wang dated 7th December 1967 – see Wang (1974), p. 8.

theories are essentially incomplete (provided they are consistent) (cf. Gödel 1931). It is interesting to see how Gödel arrived at this result.

Gödel himself wrote on his discovery in a draft reply to a letter dated 27th May 1970 from Yossef Balas, then a student at the University of Northern Iowa (cf. Wang 1987, pp. 84–85). Gödel indicated there that it was precisely his recognition of the contrast between the formal definability of provability and the formal undefinability of truth that led him to his discovery of incompleteness. One finds also there the following statement:

> [...] long before, I had found the correct solution of the semantic paradoxes in the fact that truth in a language cannot be defined in itself.

On the base of this quotation, one can argue that Gödel obtained the result on the undefinability of truth independently of A. Tarski (cf. Tarski 1933).[17]

Note also that Gödel was convinced of the objectivity of the concept of mathematical truth. In a letter to Hao Wang (cf. Wang 1974, p. 9) he wrote:

> [...] it should be noted that the heuristic principle of my construction of undecidable number-theoretical propositions in the formal systems of mathematics is the highly transfinite concept of 'objective mathematical truth' as opposed to that of 'demonstrability'.

In this situation, one should ask why Gödel did not mention the undefinability of truth, in his writings. In fact, Gödel even avoided the terms "true" and "truth" as well as the very concept of being true (he used the term "richtige Formel" and not the term "wahre Formel"). In the paper "Über formal unentscheidbare Sätze" (1931) the concept of a true formula occurs only at the end of Section 1 where Gödel explains the main idea of the proof of the first incompleteness theorem (but again the term "inhaltlich richtige Formel" and not the term "wahre Formel" appears here). Indeed, talking about the construction of a formula which should express its own unprovability invokes the interpretation of the formal system.

On the other hand, the term "truth" occurred in Gödel's lectures on the incompleteness theorems at the Institute for Advanced Study in Princeton in the spring of 1934. He discussed there, among other things, the relation between the existence of undecidable propositions and the possibility of defining the concept "true (false) sentence" of a given language in the language itself. Considering the relation of his arguments to the paradoxes, in particular to the paradox of "The Liar", Gödel indicates that the paradox disappears when one notes that the notion "false statement in a language B" cannot be expressed in B. Even more, "the paradox can be considered as a proof that 'false statement in B' cannot be expressed in B".

---

17 For the problem of the priority of proving the undefinability of truth, see Woleński (1991) and Murawski (1998).

What were the reasons of avoiding the concept of truth by Gödel? An answer can be found in a crossed-out passage of a draft of Gödel's reply to the letter of the student Yossef Balas (mentioned already above). Gödel wrote there:

> However in consequence of the philosophical prejudices of our times 1. nobody was looking for a relative consistency proof because [it] was considered axiomatic that a consistency proof must be finitary in order to make sense, 2. a concept of objective mathematical truth as opposed to demonstrability was viewed with greatest suspicion and widely rejected as meaningless.

Hence, it leads us to the conclusion formulated by S. Feferman in 1984 in the following way:

> [...] Gödel feared that work assuming such a concept [i.e., the concept of mathematical truth – my remark, R.M.] would be rejected by the foundational establishment, dominated as it was by Hilbert's ideas. Thus he sought to extract results from it which would make perfectly good sense even to those who eschewed all non-finitary methods in mathematics.

Though Gödel tried to avoid concepts not accepted by the foundational establishment, his own philosophy of mathematics was in fact Platonist. He was convinced that (cf. Wang 1996, p. 83):

> It was the anti-Platonic prejudice which prevented people from getting my results. This fact is a clear proof that the prejudice is a mistake.

Gödel's theorem on the completeness of first-order logic and his discovery of the incompleteness phenomenon together with the undefinability of truth vs. definability of formal demonstrability showed that formal provability cannot be treated as an analysis of truth, that the former is in fact weaker than the latter. It was also shown in this way that Hilbert's dreams to justify classical mathematics by means of finitistic methods cannot be fully realized. Those results together with Tarski's definition of truth (in the structure) and Carnap's work on the syntax of a language led also to the establishing of syntax and semantics in the 1930s.

On the other hand, it should be added that Gödel shared Hilbert's "rationalistic optimism" (to use Hao Wang's term) insofar as informal proofs were concerned. In fact, Gödel retained the idea of mathematics as a system of truth, which is complete in the sense that "every precisely formulated yes-or-no question in mathematics must have a clear-cut answer" (cf. Gödel 1970). He rejected however – in the light of his incompleteness theorem – the idea that the basis of these truths is their derivability from axioms. In his Gibbs lecture of 1951, Gödel distinguishes between the system of all true mathematical propositions from that of all demonstrable mathematical propositions, calling them, respectively, mathematics in the objective and subjective sense. He claimed also that it is objective mathematics that no axiom system can fully comprise.

Gödel's incompleteness theorems and in particular his recognition of the undefinability of the concept of truth indicated a certain gap in Hilbert's program and showed in particular, roughly speaking, that (full) truth cannot be comprised by provability and, generally, by syntactic means. The former can be only approximated by the latter. Hence there arose a problem: How should Hilbert's finitistic point of view be extended?

Hilbert in his lecture in Hamburg in December 1930 (cf. Hilbert 1931) proposed to admit a new rule of inference. This rule was similar to the ω-rule, but it had rather informal character (a system obtained by admitting it would be semiformal). In fact, Hilbert proposed that whenever $A(z)$ is a quantifier-free formula for which it can be shown (finitarily) that $A(z)$ is a correct (*richtig*) numerical formula for each particular numerical instance z, then its universal generalization $\forall x A(x)$ may be taken as a new premise (*Ausgangsformel*) in all further proofs.

Gödel pointed in many places that new axioms are needed to settle both undecidable arithmetical and set-theoretic propositions. In 1931 (p. 35), he stated that "[...] there are number-theoretic problems that cannot be solved with number-theoretic, but only with analytic or, respectively, set-theoretic methods". And in 1933 (p. 48) he wrote: "there are arithmetic propositions which cannot be proved even by analysis but only by methods involving extremely large infinite cardinals and similar things". In (1970) Gödel proposed "cultivating (deepening) knowledge of the abstract concepts themselves which lead to the setting up of these mechanical systems". In (1972) (this paper was a revised and expanded English version of 1958), Gödel claimed that concrete finitary methods are insufficient to prove the consistency of elementary number theory and some abstract concepts must be used in addition. In the paper (1946), Gödel explicitly called for an effort to use progressively more powerful transfinite theories to derive new arithmetical theorems.

Also Zermelo proposed to allow infinitary methods to overcome restrictions revealed by Gödel. According to Zermelo, the existence of undecidable propositions was a consequence of the restriction of the notion of proof to finitistic methods (he said here about "finitistic prejudice"). This situation could be changed if one used a more general "scheme" of proof. Zermelo had here in mind an infinitary logic, in which there were infinitely long sentences and rules of inference with infinitely many premises. In such a logic, he insisted, "all propositions are decidable!" He thought of quantifiers as infinitary conjunctions or disjunctions of unrestricted cardinality and conceived of proofs not as formal deductions from given axioms but as metamathematical determinations of the truth or falsity of a proposition. Thus syntactic considerations played no role in his thinking.

To give a rough account of how those suggestions and proposals to extend the finitistic point of view do in fact work, let us quote some technical results. We restrict ourselves to the case of the arithmetic of natural numbers, more exactly to Peano arithmetic PA.

Generally speaking, one can obtain completions of PA by:

- admitting the ω-rule,
- adding new axioms (in particular reflection principles) and
- adding (partial) notion(s) of truth.

Let us start by considering the case of the ω-rule, i.e., of the following rule:

$$\frac{\varphi(\overline{0}), \varphi(\overline{1}), \varphi(\overline{2}), \ldots \quad (n \in \mathbb{N})}{\forall x \varphi(x)}.$$

Denote by $(PA)_\omega$ Peano arithmetic PA with the ω-rule. One can easily see that $(PA)_\omega$ is complete – it follows from the fact that its unique model up to isomorphism is the standard model $\mathcal{N}_0 = \langle \mathbb{N}, S, 0, +, \cdot \rangle$. Hence $(PA)_\omega = \text{Th}(\mathcal{N}_0)$.

One can ask: How many times must the ω-rule be applied to obtain a complete extension of PA? To give an answer, let us define the following hierarchy of theories where T is any first-order theory in the language L(PA) of Peano arithmetic:

$$T^0 = T,$$

$$T^{\alpha+\frac{1}{2}} = T^\alpha \cup \{\varphi : \varphi \text{ is of the form } \forall x \psi(x) \text{ and } \psi(\overline{n}) \in T^\alpha \text{ for every } n \in \mathbb{N}\},$$

$$T^{\alpha+1} = \text{the smallest set of formulas containing } T^{\alpha+\frac{1}{2}} \text{ and closed under the rules of inference of PA},$$

$$T^\lambda = \bigcup_{\alpha < \lambda} T^\alpha \text{ for } \lambda \text{ limit}.$$

One can now prove that

Theorem 1 $\text{Th}(\mathcal{N}_0) = (PA)_\omega = PA^\omega$.

Recall the hierarchy of formulas of the language L(PA). Let $\Sigma_0^0 = \Pi_0^0 = \Delta_0^0$ be the set of all quantifier free formulas and all formulas with bounded quantifiers. Define Σ_{n+1}^0 to be the set of all formulas of the form $\exists x \psi$ for $\psi \in \Pi_n^0$, and Π_{n+1}^0 to be the set of all formulas of the form $\forall x \psi$ for $\psi \in \Sigma_n^0$. We also define Δ_n^0 as the set of all formulas equivalent (in PA) to a Σ_n^0 formula and to a Π_n^0 formula. One can prove that

Theorem 2 *For every $n \in \mathbb{N}$ the theory PA^n is complete with respect to Σ_{2n+1}^0 sentences.*

In PA one can define partial notions of truth, i.e., one can define satisfaction and truth for formulas of a given class of the arithmetical hierarchy. Denote by $Sat_{\Sigma_n^0}$

the definition of satisfaction for Σ_n^0 formulas; similarly let $Sat_{\Pi_n^0}$ denote the definition of satisfaction for Π_n^0 formulas. Note that $Sat_{\Sigma_0^0}$ and $Sat_{\Pi_0^0}$ are Σ_1^0 formulas and that $Sat_{\Sigma_n^0}$ and $Sat_{\Pi_n^0}$ (for $n \geq 1$) are Σ_n^0 and Π_n^0, respectively. Let further $Tr_{\Sigma_n^0}$ and $Tr_{\Pi_n^0}$ denote truth predicates for Σ_n^0 and Π_n^0 sentences.[18] In the sequel, we shall identify formulas defining satisfaction and truth and their extensions in the standard model $\mathcal{N}_0$.

The previous theorem can now be formulated as:

$$\text{PA}^n \supseteq \text{PA} + Tr_{\Sigma_{2n+1}^0}.$$

In the definition of the hierarchy T^α no restriction was put on formulas to which the ω-rule was applied. Consider now a hierarchy in which such a restriction is put. So let T be any theory in the language L(PA). Define the following hierarchy of theories (cf. Niebergall 1996):

$T^{(0)} \quad = \quad T,$

$T^{(\alpha + \frac{1}{2})} \quad = \quad T^{(\alpha)} \cup \{\varphi : \varphi \text{ is of the form } \forall x \psi(x) \text{ and } \psi(x) \in \Sigma_{2\alpha+1}^0 \text{ and } \psi(\overline{n}) \in T^{(\alpha)} \text{ for every } n \in \mathbb{N}\},$

$T^{(\alpha+1)} \quad = \quad$ the smallest set of formulas containing $T^{(\alpha + \frac{1}{2})}$ and closed under the rules of inference of PA,

$T^{(\lambda)} \quad = \quad \bigcup_{\alpha < \lambda} T^{(\alpha)}$ for λ limit.

Hence the ω-rule is now applied at stage n to Σ_{2n+1}^0 formulas only.

One has the following

Theorem 3 (Niebergall 1996) *For any $n \in \mathbb{N}$,*

$$\text{PA}^{(n)} = \text{PA} + Tr_{\Sigma_{2n+1}^0}.$$

The above theorems[19] indicate interconnections between Peano arithmetic augmented with the ω-rule and the partial truths. Other connections between them can be formulated in the language of interpretability. So let $S \preccurlyeq T$ denote that a theory S is relatively interpretable in the theory T (in the sense of Tarski 1953). We have now the following facts (cf. Niebergall 1996):

---

18 Construction of $Sat_{\Sigma_n^0}$ and $Sat_{\Pi_n^0}$ can be found in Kaye (1991) and Murawski (1999).
19 Note that many of those theorems hold not only for Peano arithmetic PA but also for a broad class of theories – cf. Niebergall (1996).

Theorem 4 *Let* Con_S *(for an appropriate theory S) denote a statement of* L(PA) *stating that S is consistent. Then*

$$PA^n \preccurlyeq PA + Tr_{\Sigma^0_{2n+1}} + Con_{PA^n},$$

$$PA^n \preccurlyeq PA + Tr_{\Sigma^0_{2n+2}},$$

$$PA + Con_{PA + Tr_{\Sigma^0_n}} \preccurlyeq PA + Tr_{\Sigma^0_2},$$

$$PA + Con_{PA^n} \preccurlyeq PA + Tr_{\Sigma^0_2}.$$

The above theorems indicate that the arithmetical truth, i.e., the set $Th(\mathcal{N}_0)$ of all arithmetical sentences true in the standard model $\mathcal{N}_0$, can be approximated by syntactical methods, i.e. by demonstrability – though not by finitary means (one uses here the ω-rule).

So far, we have considered Peano arithmetic and partial truths. Ask now: What about PA and the full truth? Gödel's and Tarski's theorem shows that the truth predicate for L(PA) cannot be defined in PA. But one can extend the language L(PA) by adding a new binary predicate S called satisfaction class and characterizing it axiomatically by adding to Peano arithmetic PA axioms being an appropriate modification of Tarski's definition of satisfaction (cf. Krajewski 1976, where this notion was introduced, or Murawski, 1997). Note that since those axioms form a finite set of axioms, one can write them as a single formula of the language L(PA) ∪ S. Denote this theory as PA + "S is a satisfaction class". One can extend this theory by adding new axioms stating special properties of S. In particular, one can demand that S is full, i.e., S decides any formula of L(PA) on any valuation or that S is Γ-inductive for Γ being a given class of formulas of the language L(PA) ∪ S, i.e., that the induction axiom holds for all formulas of the class Γ (if Γ is the class of all formulas of L(PA) ∪ S then one says that the satisfaction class S is inductive).

Since theories T of the indicated type are extensions of PA one can ask what about natural numbers can be proved in T, i.e., one can consider theories of the type

$$PA^T = \{\varphi \in L(PA) : T \vdash \varphi\}.$$

Theorems of PA^T are those sentences of the language L(PA) of Peano arithmetic (hence sentences about natural numbers) which can be proved in the stronger theory T. A natural problem of finding an axiomatization of the theory PA^T arises.

One can easily see that the following theories are conservative extensions of PA:

(a) PA + "S is a satisfaction class",
(b) PA + "S is a full satisfaction class" and
(c) PA + "S is an inductive satisfaction class".

This means that one can prove in those theories exactly the same theorems about natural numbers (i.e., formulas of the language L(PA)) as in Peano arithmetic PA. Hence the addition of a new notion, i.e., of a notion of a satisfaction class (and consequently a notion of truth), with properties indicated in (a)–(c) does not increase the proof-theoretical power of a theory with respect to sentences of the language L(PA). On the other hand, the assumption that a satisfaction class is full and Δ_0^0-inductive gives a nonconservative extension of PA. In fact one can prove in this theory the consistency of PA.

The theories PA^T for T being PA + "S is a full (Σ_m^0-)inductive satisfaction class" can be characterized by transfinite induction or the consistency of appropriate ω-logics. Denote by $\Gamma - \text{PA}(S)$ the theory PA + "S is a full Γ-inductive satisfaction class" and by PA(S) the theory PA + "S is a full inductive satisfaction class".

Consider the following sequence of formulas of the language L(PA) (one uses here arithmetization):

$$\Gamma_0(\varphi) = \text{``PA} \vdash \varphi\text{''},$$

$$\Gamma_{n+\frac{1}{2}}(\varphi) = \text{``}\varphi \text{ is of the form } \eta \vee \forall z\, \psi(z) \text{ and } \forall z\, \Gamma_n(\eta \vee \psi(S^z 0))\text{''},$$

$$\Gamma_{n+1}(\varphi) = \text{``there exists a proof of the formula } \varphi$$
$$\text{based on } \text{PA} \cup \{\psi : \Gamma_{n+\frac{1}{2}}(\psi)\}\text{''}.$$

Observe that in this system of ω-logic only the application of the ω-rule increases the degree of complexity of a proof.

Theorem 5 (Kotlarski 1986) $\text{PA}^{\Delta_0^0 - \text{PA}(S)} = \text{PA} \cup \{\neg\Gamma_n(0 = 1) : n \in \mathbb{N}\}.$

It can also be proved (cf. Kotlarski 1986) that the theory $\Delta_0^0 - \text{PA}(S)$ is equal to the theory

$$\text{PA} + S \text{ is a full satisfaction class} + \forall\varphi[(\text{PA} \vdash \varphi) \to S(\varphi)].$$

The last sentence can be read as: "S makes all theorems of PA true". It is equivalent to the Δ_0^0-inductiveness of the satisfaction class S.

The system of ω-logic described above can be iterated in the transfinite and one can axiomatize theories $\text{PA}^{\Sigma_m^0 - \text{PA}(S)}$ ($m \in \mathbb{N}$) and $\text{PA}^{\text{PA}(S)}$ by consistency statements of appropriate systems of this logic (cf. Kotlarski and Ratajczyk 1990a).

Define for an ordinal α a sequence $\omega_m(\alpha)$ in the following way: $\omega_0(\alpha) = \alpha$, $\omega_{m+1}(\alpha) = \omega^{\omega_m(\alpha)}$ and put $\omega_n = \omega_n(\omega)$. Let now $TI(\rho)$, where ρ is an ordinal, denote the scheme of transfinite induction up to ρ. Then the following theorem holds.

Theorem 6 (Kotlarski and Ratajczyk 1990b) *Let m be a natural number. Then*

$$PA^{\Sigma_m^0-PA(S)} = PA \cup \{TI(\varepsilon_{\omega_m(k)}) : k \in \mathbb{N}\}.$$

$$PA^{PA(S)} = PA \cup \{TI(\varepsilon_{\omega_k}) : k \in \mathbb{N}\}.$$

The above theorems show how strong is Peano arithmetic augmented with an appropriate notion of satisfaction (and truth). One can see that only by assuming that the added notion of satisfaction (truth) is full and at least Δ_0^0-inductive one obtains a proper extension of PA. It is interesting that such extensions are equivalent to PA extended by appropriate forms of transfinite induction or by the statements of the consistency of appropriate systems of ω-logic. In other words, the above theorems show in particular that what can be proved about natural numbers using Peano axioms and the notion of satisfaction (truth) that is assumed to be full and Σ_m^0-inductive is exactly the same as what can be proved in PA plus transfinite induction for ordinals $\varepsilon_{\omega_m(k)}$ (for all $k \in \mathbb{N}$) or in PA plus appropriate consistency statements. Similarly for PA plus full inductive satisfaction (truth) on the one hand and PA plus transfinite induction for ordinals ε_{ω_k} (for all $k \in \mathbb{N}$) or PA plus appropriate consistency statements on the other. They show also that by adding to PA the notion of satisfaction (truth) and assuming that it is full and makes all theorems of PA true, one obtains a theory with exactly the same theorems about natural numbers as by taking PA augmented with a concept of a full and Λ_0^0-inductive satisfaction (truth) or PA plus appropriate consistency statements.

In the above considerations, we restricted ourselves to formal proofs and to the semantical notion of truth *in* mathematics. We tried to show how the awareness of differences between them has been developed – from the hopes that formal proofs provide sufficient means to exhaust the mathematical truth to the discovery of various limitations of them. Let us finish with some general remarks.

Concepts of proof and truth are (even in mathematics) ambiguous. One should distinguish between working proofs of everyday mathematics and idealized formal proofs used by logicians. On the other hand, a proof in mathematics has various aspects and can be studied from various points of view. One can distinguish psychological, social, cultural and logical aspects of proofs. a proof can be studied as a mathematical or as an epistemological object. The former is precisely defined on the basis of mathematical logic, and the latter is a vague concept. The former is an idealization of proofs occurring in a research practice of mathematicians, is a reconstruction of them. Recently, one can observe in the philosophy of mathematics a tendency to concentrate on the actual research practice of mathematicians rather than on idealized foundational reconstructions of it and consequently to study the methods actually used by mathematicians.

Similar distinctions can be made with respect to the concept of truth. The semantical concept of truth precisely defined by Tarski is in fact a mathematical notion. It provides a definition of truth *in* mathematics;[20] it is the concept of truth for a model in a formal language (its essential feature is to define truth in terms of reference or satisfaction on the basis of a particular kind of syntactico-semantical analysis of the language). But one can also speak about epistemic truth – cf., e.g. Isaacson (1987, 1992) where it is argued that Peano arithmetic is complete with respect to an epistemic notion of arithmetical truth.

The distinction between proof and truth in mathematics presupposes of course some philosophical assumptions. In fact for pure formalists and for intuitionists there exists no truth/proof problem. For them a mathematical statement is true just in case it is provable, and proofs are syntactic or mental constructions of our own making. In the case of a Platonist (realist) philosophy of mathematics, the situation is different. One can say that Platonist approach to mathematics enabled Gödel to state the problem and to be able to distinguish between proof and truth, between syntax and semantics.[21]

---

20 One should distinguish the truth *in* mathematics and the truth *of* mathematics.
21 Note that, as indicated above, Hilbert was not interested in philosophical questions and did not consider them.

Some Historical, Philosophical and Methodological Remarks on Proof in Mathematics

Introduction

Proofs play an important role in mathematics and its methodology (in the context of justification). They form the main method of justifying mathematical statements. Only statements that have been proved can be treated as belonging to the corpus of mathematical knowledge. Proofs are used to convince the readers of the truth of presented theorems. But what is in fact a proof? In mathematical research practice, proof is a sequence of arguments that should show the truth of the claim. Of course, the particular arguments used in a proof depend on the situation, on the audience, on the type of a claim, etc. Hence a concept of a proof has in fact a cultural, psychological and historical character. In practice, mathematicians generally agree whether a given argumentation is or is not a proof. More difficult is the task to define a proof as such. Beside proofs used in the research practice there is a concept of a formal proof developed by logic. What are the relations between them? What roles do they play in mathematics?

Problems of that type will be considered in the paper. We start by some historical remarks showing in what circumstance the idea of a proof (informal and formal) appeared. Next, the features and role played by informal proofs will be considered. The subject of the next section will be formal proofs and their relation to the concept of truth. In the closing section, some conclusions will be made and a thesis (similar to Church-Turing Thesis in the computation theory) formulated.

Historical remarks

The model of mathematics as a science, its paradigm functioning in fact till today, was formulated and developed in the ancient Greece about 4th century B.C. Earlier, e.g. in ancient Egypt or Babylon, mathematics consisted of practical procedures that should help to solve everyday problems such as measuring surface area or the amount of cereal in a granary or oil in a jug. In those both pre- Greek mathematics – though they were advanced and sophisticated (especially the Babylonian mathematics) – there was no need to prove statements. In fact, there were no general statements and no attempts were undertaken to deduce the results or to explain their validity. One was satisfied by instructions what should be done in order to receive the result or to perform the required task. In fact, mathematics there was a collection of separate algorithms (as one would say today) and resembled in certain sense informatics (though without sophisticated technical equipment). Similar was the situation in China – Chinese mathematics was also a collection of procedures (transferred from generation to generation).

Proofs as deduction from explicitly stated postulates was conceived by the Greeks. It was connected with the axiomatic method. Since Plato, Aristotle and Euclid the axiomatic method was considered as the best method to justify and to organize mathematical knowledge. The first mature and most representative example of its usage in mathematics was the *Elements* of Euclid. They established a pattern of a scientific theory and in particular a paradigm in mathematics. Since Euclid till the end of the 19th century, mathematics was developed as an axiomatic – in fact rather a quasi-axiomatic – theory based on axioms, postulates and definitions. Axioms were principles common to all sciences, postulates – specific principles taken for granted by a mathematician engaged in the demonstration of theorems in a particular domain. Definitions should provide meaning to new notions – in practice definitions were rather explanations of notions than proper definitions in the strict sense, moreover, they were explanations in the unprecise everyday colloquial language. Note that the language of a theory was not separated from the natural language. Proofs of theorems contained several gaps – in fact the lists of axioms and postulates were not complete; one freely used in proofs various "obvious" truths or referred to the intuition. Consequently, proofs were only partially based on axioms and postulates. In fact proofs were informal and intuitive, they were rather demonstrations and the very concept of a proof was of a psychological and sociological (and not of a logical) nature.

Note that the language of theories was simply the unprecise colloquial language. Till the end of the 19th century, mathematicians were convinced that axioms and postulates should be true statements, hence sentences describing the real state of affairs (in the mathematical reality).[22] It seems to be connected with Aristotle's view that a proposition is demonstrated (proved to be true) by showing that it is a logical consequence of propositions already known to be true. Demonstration was conceived here of as a deduction whose premises are known to be true, and a deduction was conceived of as a chaining of immediate inferences.

It should be noted that Euclid's approach (connected with Platonic idealism) to the problem of the development of mathematics and the justification of its statements (which found its fulfilment in the Euclidean paradigm), i.e. justification by deduction (by proofs) from explicitly stated axioms and postulates, was not the only approach and method which was used in the ancient Greek (and later). The

---

22 In fact, the discussion about the truth of mathematical statements started already with the discovery of non-Euclidean geometries, but the conviction of the truth of mathematical theses dominated.

other one (call it heuristic) was connected with Democritus' materialism. It was applied, e.g., by Archimedes who used in his mathematical works not only deduction but any methods, such as intuition or even experiments (not only mental ones) to solve problems. One can see this, e.g., in his considerations concerning the calculation of the volume of a sphere using cylinder with two excavated cones or in his quadrature of the parable.

Though the Euclidean approach won and dominated in the history, one should note that it formed rather an ideal and not the real scientific practice of mathematicians. In fact rigorous, deductive mathematics was rather a rare phenomenon. On the contrary, intuition and heuristic reasoning were the animating forces of mathematical research practice. The vigorous but rarely rigorous mathematical activity produced "crises" (e.g. the Pythagoreans' discovery of the incommensurability of the diagonal and the side of a square, Leibniz's and Newton's problems with the explanation of the nature of infinitesimals, Fourier's "proof" that any function is representable in a Fourier series, antinomies connected with Cantor's imprecise and intuitive notion of a set).

New elements appeared in the 19th century with the trend whose aim was the clarification of basic mathematical concepts, especially those of analysis (cf. works by Cauchy, Weierstrass, Bolzano, Dedekind). Still another factor was the discovery of antinomies, in particular in set theory (C. Burali-Forte, G. Cantor, B. Russell) and of the semantical antinomies (G.D. Berry, K. Grelling). They forced the revision of some basic ideas of mathematics. Among formulated proposals was the foundational program of David Hilbert and his *Beweistheorie*. Note that "this program was never intended as a comprehensive philosophy of mathematics; its purpose was instead to legitimate the entire corpus of mathematical knowledge" (Rowe 1989, p. 200).

The main role in Hilbert's program was played by formal (or formalized) proofs. This device was introduced on the basis of (and thanks to) the mathematical logic developed in the 19th century, in particular on the basis of the work of Gottlob Frege who constructed the first formalized system – it was the system of propositional calculus based on two connectives: negation and implication. Investigations carried out in the framework of Hilbert's program established the new scientific discipline, i.e. the metamathematics.

Recall that the concept of a formal proof must be related to a given formal theory. To express such a theory one should at the beginning fix its language. The rules of forming formulas in it should have strictly formal and syntactic character – only the shape of symbols can be taken into account and one should entirely abstract from their possible meaning or interpretation. Next one fixes axioms (logical, non-logical and usually identity axioms) and rules of inference. The latter must have entirely syntactic and formal character. A formal proof of a statement (formula) φ

is now a finite sequence of formulas in the given language $\varphi_1, \varphi_2, \varphi_3, \ldots, \varphi_n$ such that the last member of the sequence is the formula φ, and all members of it either belong to the set of presumed axioms or are consequences of previous members of the sequence according to one of the accepted rules of inference. Observe that this concept of a formal proof has a syntactic character and does not refer to any semantical notions such as meaning or interpretation.

As a result of the described development, one has to deal nowadays in mathematics with at least two concepts of a proof: the formal one, used mainly by logicians and specialists in metamathematics in the foundational studies as well as by computer scientists (cf. modern theorem provers), and on the other hand the "normal", "usual" concept of a proof used in mathematical research practice. What are the relations between them? Is the metamathematical concept a precise explication of the "everyday" concept? Do they play similar roles in mathematics or not? To what extent does the concept of a formal proof reflect the main features and the nature of an informal proof used by mathematicians in their researches?

Informal proofs and their role

Mathematics was and is developed in an informal way using intuition and heuristic reasonings – it is still developed in fact in the spirit of Euclid (or sometimes of Archimedes) in a quasi-axiomatic way. Moreover, informal reasonings appear not only in the context of discovery but also in the context of justification. Any correct methods are allowed to justify statements. But what does it mean "correct"? In the research practice, this was and is decided by the community of mathematicians. Consequently, the criteria of being correct have been changing in the history and in the process of developing mathematics. The concept of proof was and is in fact not an absolute notion but it was and is culturally and sociologically dependent and motivated; it had and has cultural and sociological components. The main aim of a mathematician is always to convince the audience that the given result is justified, correct and true (the latter concept being used in an intuitive and vague way) and not to answer the question whether it can be deduced from stated axioms. In the research practice, a proof is in fact an argumentation that should indicate the correctness of a claimed thesis – in particular its form depends on and is relative to a background knowledge of those to whom the proof is presented. And here appears a psychological moment in the understanding and functioning of proofs.

The ultimate aim of mathematics is "to provide correct proofs of true theorems" (Avigad 2006, p. 105). In their research practice, mathematicians usually do not distinguish concepts "true" and "provable" and often replace them by each other. Mathematicians used to say that a given theorem holds or that it is true and not that it is provable in such and such theory. In fact, they do not distinguish concepts "true" and "provable" and often replace them by each other, Add that axioms of theories being developed are not always precisely formulated and admissible methods precisely described.

Proofs play various roles in the mathematical research practice. One can distinguish (cf. CadwalladerOlsker 2011; De Villiers 1999) among others the role of: (1) verification, (2) explanation, (3) systematization, (4) discovery, (5) intellectual challenge, (6) communication and (7) justification of definitions.

The most familiar to research mathematicians is the role of verification. A statement can be treated as belonging to the body of mathematics only when it has been verified. The proof should not only show that a given sentence is true and holds but should also explain why it is true and holds. This role explains why mathematicians are often looking for new proofs of known theorems – new proofs should have more explanatory power. The role of systematization was exemplified already by Euclid's *Elements*. In this work, many theorems known to Greeks have been collected and organized in such a way that they followed from axioms, postulates, definitions and previously proved theorems. It was shown in this way that the accepted axioms, postulates and definitions form a sufficient base on which the whole edifice of mathematics can be developed. Note that the role of discovery may be – *prima facie* – rather seldom associated with proofs but it is not excluded. In fact, e.g. non-Euclidean geometries were arrived at through purely deductive means. Recall that since Euclid one asked the question whether the fifth postulate on parallels formulated in *Elements* is independent of other axioms and postulates or can be deduced from them. After several attempts undertaken through the centuries, it has been shown in the 19th century that it is reasonable to consider systems of geometry in which the negation of the fifth postulate is assumed instead of the fifth postulate itself, and it is possible to show that such systems are consistent.

Finding proofs is the intellectual challenge for mathematicians: there is a theorem – we want to prove it. Proofs serve in the community of mathematicians as communication means. They communicate not only the reasons why a given statement is a true theorem but introduce also new methods which can be used sometimes in other domains. Proofs can provide also justification of definitions.

The most important roles played by proofs in the research mathematical practice are of course verification and explanation. Note that a proof that verifies a theorem does not have to explain why it holds. One can distinguish between proofs that convince and proofs that explain. The former should show that a statement holds or is true and can be accepted, the latter – why it is so. Of course, there are proofs that both convince and explain. The explanatory proof should give an insight in the matter whereas the convincing one should be concise or general. One can distinguish also between explanation and understanding. Mathematicians often treat simplicity as a characteristic feature of understanding. Observe that, as G.-C. Rota writes in (1997, p. 192):

> [i]t is an article of faith among mathematicians that after a new theorem is discovered, other, simpler proof of it will be given until a definitive proof is found.

In this context, it is worth distinguishing between unveiling proofs and consolidating ones. The former one is "a proof which proves a theorem which was unknown before" and the latter is "a proof of a theorem which is already known to be true" (Kahle 2015). Proofs of the second type do not contribute to the truth of a theorem – they consolidate our knowledge of this truth. Such a proof can be quite different from the original one. Aschbacher wrote about this phenomenon in the following way (2005, p. 2403):

> The first proof of a theorem is usually relatively complicated and unpleasant. But if the result is sufficiently important, new approaches replace and refine the original proof, usually by embedding it in a more sophisticated conceptual context, until the theorem eventually comes to be viewed as an obvious corollary of a larger theoretical construct. Thus proofs are a means for establishing what is real and what is not, but also a vehicle for arriving at a deeper understanding of mathematical reality.

There are many examples from the history of mathematics that confirm this. In this context, one can mention Paul Erdős' idea of proofs from The Book in which God maintains the perfect proofs for mathematical theorems – he followed the dictum of G.H. Hardy that there is no permanent place for ugly mathematics. Erdős stressed that one need not believe in God but, as a mathematician, one should believe in The Book.[23]

Note that a proof that convinces can be more (or even quite) formal. Explanatory proofs cannot be strictly formal. Mathematicians set a high value on explanatory proofs. Such a proof is more valued when "it exhibits methods that are powerful and informative" (Avigad 2006, p. 106). Hersh says in (1997, p. 60) that "[p]roof is a tool in service of research, not a shackle on the mathematician's imagination".[24]

Sometimes theorems are verified in mathematics by checking all particular cases, but this usually does not give an explanation why the theorem holds. The explanation should give a general principle by which the theorem holds. a famous example of a theorem verified by checking cases but not giving reasons is the Four-Color Theorem proved by Appel and Haken and stating that every planar graph is four-colourable, i.e. in another words, that four colours suffice to colour every map

---

23 In 1998, Springer Verlag published *Proofs From The Book* by M. Aigner and G.M. Ziegler (cf. Aigner and Ziegler 1998). It is treated by the authors as "a first (and very modest) approximation of The Book" (Preface).

24 On the concept of explanation in mathematics see, e.g. Steiner (1978) and Mancosu (2000, 2001).

on the plane in such a way that two regions receive different colours whenever they have a common border.

The example of the Four-Color Theorem indicates other features of proofs in mathematics. Observe that the first purported proof of it given by Kempe in 1879 was accepted for a decade before it was found to be incorrect. This was neither the first nor the last example of such a situation. It means that the community of mathematicians can be fallible.

The Four-Color Theorem opened eyes of philosophers of mathematics to the problem of methods acceptable in a proof or in a verification of cases. The unique known proof – that by Appel and Haken – was obtained by using computer, and no traditional proof (without computer) is known so far. Moreover, the existing proof cannot be made by a human being because an essential part of it was a computation requiring about 1200 hours of computer time and is beyond the capacities of any mathematician. This initiated a discussion concerning the admissibility of experimental methods in mathematical proofs. Several arguments pro and contra have been formulated – we will not enter here the details of the discussion. Let us say only that the usage of a technical tool (like a computer) seems to refute the commonly accepted thesis that mathematical knowledge is *a priori*. There arises also a question whether a computer-aided proof is (or can be treated as) a mathematical proof and consequently, whether in particular the Four-Color Theorem can be called "theorem" or it is still rather a hypothesis.

In this debate initiated by a paper by Tymoczko (1979) a question was asked what are in fact the characteristic features of a "normal" mathematical proof. Tymoczko says that a proof in mathematics should be: (1) convincing, (2) surveyable and (3) formalizable.

The first feature is – as Tymoczko says – of an "anthropological" character, the other two are treated by him as "deep features". He claims also that "surveyability and formalizability can be seen as two sides of the same coin" (Tymoczko 1979, p. 61). Formalizability "idealizes surveyability, analyzes it into finite reiterations of surveyable patters" (*ibidem*). It can be assumed that all surveyable proofs are formalizable. Are also all formalizable proofs surveyable? Tymoczko answers this negatively: "[w]e know that there must exist formal proofs that cannot be surveyed by mathematicians if only because the proofs are too long or involve formulas that are too long" and the phrase "too long" means here "can't be read over by a mathematician in a human life time" (*ibidem*). On the other hand, one should observe that "it is not at all obvious that mathematicians could come across formal proofs and recognize them as such without being able to survey them" (Tymoczko 1979, p. 62).

Considering surveyability one should distinguish local and global surveyability. Bassler in (2006, p. 100) characterizes them in the following way:

local surveyability requires the surveying of each of the individual steps in a proof in some order, while global surveyability requires the surveying of the entire proof as a comprehensive whole.

Hence, a local surveyability does not mean that a proof is practically surveyable. One can say that the proof of the Four-Color Theorem is globally surveyable without being locally surveyable (provided one is willing to countenance a distinction between proof and calculation). On the other hand, if one accepts the assumption that global surveyability receives its foundation in local surveyability then this statement is false. Add that one should also distinguish the surveyability of a proof and the fact that it can be (formally) checked on the one hand and the fact that it gives an understanding, that it reveals the deep reasons for the theorem being proved on the other.

The concept of surveyability is not precise enough. In the 20th century, there was a trend to link surveyability with the development of formal and complete foundations of mathematics, and formalization was treated as a method providing the local surveyability. The works of Frege, Russell and their followers, especially Hilbert, were guided by the desire to find a perspicuous syntactic representation of the relations of semantic content within a proposition.

Computers and methods connected with them were and are used in mathematics not only in the proof of the Four-Color Theorem. They are used in various contexts in mathematics, in particular: (1) to perform numerical calculations, (2) to find (usually approximate) solutions of equations and systems of algebraic or differential equations or of integrals, (3) in automating proofs of theorems, (4) in checking the correctness of mathematical proofs, (5) in proving theorems (one says then about computer-aided proofs) and (6) in various experiments with mathematical objects (e.g., in the theory of fractals).

From our point of view, the most important applications are (3) and (4) – the application of the type (5) has been discussed above on the example of the Four-Color Theorem.

The automating proving of theorems is connected with the idea of mechanization and automatization of reasoning due to Leibniz (cf. Marciszewski and Murawski 1995). This idea (as one of the factors) led to the development of formal logic and in consequence to the idea of a formal proof.

Formal proofs and their role

Formal proofs were introduced to provide an explication of the informal notion of a proof and to solve some metamathematical problems. They should explain the virtue by which usual proofs used in the research practice are judged to be correct. They should also explain what does it mean that a given statement is a logical or deductive consequence of certain assumptions. As indicated above they were introduced in the atmosphere of a crisis in the foundations of mathematics. For Frege and Russell they were means to an end, a way of precisely isolating the permissible proofs and making sure that all use of axioms was explicit. On the other hand, observe that Hilbert was not really interested in actually formalizing proofs and replacing the "normal" research proofs by formalized ones. He treated formalization and formal proofs as a tool to justify mathematics as a science and to establish its consistency. They should serve theoretical purposes – in particular to prove results about mathematics, hence to obtain metamathematical results.

In fact, the development of logic and the concept of a formal proof based only on syntactical properties made possible the development of metamathematics. A lot of interesting results have been obtained here. First of all the old paradigm of mathematics that was functioning since Euclid has been made precise – in fact it has been replaced by a new logico-set-theoretical paradigm (cf. Batóg 1996). The main features of this new paradigm can be described as follows: (1) Set theory became the fundamental domain of mathematics, in particular some set-theoretical notions and methods are present in any mathematical theory; and set theory is the basis of mathematics in the sense that all mathematical notions can be defined by primitive notions of set theory and all theorems of mathematics can be deduced from axioms of set theory. (2) Languages of mathematical theories are strictly separated from the natural language, they are artificial languages and the meaning of their terms is described exclusively by axioms; some primitive concepts are distinguished and all other notions are defined in terms of them according to precise rules of defining notions. (3) Mathematical theories have been axiomatized. (4) There is a precise and strict distinction between a mathematical theory and its language on the one hand and metatheory and its metalanguage on the other (the distinction was explicitly made by A. Tarski).

Note also that two concepts, crucial for mathematics: the concept of a syntactical consequence (being provable) and the concept of being a semantical consequence, have been precisely defined and strictly distinguished. One could also precisely distinguish provability and truth. In a "normal" research mathematical practice – as we indicated above – they are usually identified or at least not distinguished – one says that a theorem holds, i.e. has been proved (in a "normal", informal sense of this notion), or that it is true and treats both as equivalent and synonymous. The very process of distinguishing them was long and not so simple – cf. Murawski (2002a, 2002b). The crucial role was played here by Gödel's incompleteness theorems.

The incompleteness results of Gödel showed that any theory (based on a recursive set of axioms and finitary rules of inference) including arithmetic of natural numbers is in fact incomplete, hence there exist sentences that are true but are neither provable nor refutable, i.e. they are undecidable in a given theory. Before Gödel, it was believed that formal demonstrability is an analysis of the concept of mathematical truth. Gödel wrote in a letter of 7th March 1968 to Hao Wang (cf. Wang 1974, p. 10):

> [...] formalists considered formal demonstrability to be an analysis of the concept of mathematical truth and, therefore were of course not in a position to distinguish the two.

Indeed, the informal concept of truth was not commonly accepted as a definite mathematical notion at that time.[25] Gödel wrote in a crossed-out passage of a draft of his reply to a letter of the student Yossef Balas: "[...] a concept of objective mathematical truth as opposed to demonstrability was viewed with greatest suspicion and widely rejected as meaningless" (cf. Wang 1987, pp. 84–85). It is worth comparing this with a remark of R. Carnap. He writes in his diary that when he invited A. Tarski to speak on the concept of truth at the September 1935 International Congress for Scientific Philosophy, "Tarski was very sceptical. He thought that most philosophers, even those working in modern logic, would be not only indifferent, but hostile to the explication of the concept of truth". And indeed at the Congress "[...] there was vehement opposition even on the side of our philosophical friends" (cf. Carnap 1963, pp. 61–62).

All these explains in some sense why Hilbert preferred to deal in his metamathematics solely with forms of formulas, using only finitary reasonings which were considered to be safe – contrary to semantical reasonings which were non-finitary and consequently not safe. Non-finitary reasonings in mathematics were considered to be meaningful only to the extent to which they could be interpreted or justified in terms of finitary metamathematics.[26] On the other hand, there was no clear distinction between syntax and semantics at that time. Recall, for example, that the axiom systems came by Hilbert often with a built-in interpretation. Add also that the very notions necessary to formulate properly the difference syntax–semantics were not available to Hilbert though he was aware of the complex of

---

25 Note that there was at that time no precise definition of truth – this was given only in 1933 by A. Tarski (1933).

26 Cf. Gödel's letter to Hao Wang dated 7th December 1967 – see Wang (1974, p. 9). He wrote there: "I may add that my objectivist conception of mathematics and metamathematics in general, and of transfinite reasoning in particular, was fundamental also to my other work in logic. How indeed could one think of expressing metamathematics in the mathematical systems themselves, if the latter are considered to consist of meaningless symbols which acquire some substitute of meaning only through metamathematics. [...] it should be noted that the heuristic principle of my construction of undecidable number theoretical propositions in the formal systems of mathematics is the highly transfinite concept of 'objective mathematical truth' as opposed to that of 'demonstrability' (cf. M. Davis, *The Undecidable*, New York 1965, p. 64, where I explain the heuristic argument by which I arrive at the incompleteness results), with which it was generally confused before my own and Tarski's work".

problems – cf. his statement of the question in Hilbert and Ackermann (1928) which led to Gödel's completeness theorem. Gödel proved that truth cannot be adequately achieved and expressed by provability, that the whole of mathematics (or even parts of it) cannot be included in a formalized system. This indicated certain weakness of the concept of a formal proof. Gödel's results showed also that one should not limit or bound the creative invention of mathematicians. In the framework of formalized theories, one can extend them by adding new axioms or by admitting new inference rules. The second possibility means that infinitary rules are admitted – but this changes the whole picture and the whole paradigm! Note that Hilbert had in fact no problem with such a change, to the opposite – in view of Gödel's result – he encouraged it. On the other hand, one can ask whether the process of adding new axioms, though necessary to solve problems undecidable in a given theory, is sufficient? Will it suffice to express the creativity of a mathematical mind, the creativity of mathematicians?

The incompleteness theorems of Gödel belong to the so-called limitation results. They are results stating that certain properties important and desired from a metamathematical point of view (but also from the point of view of a working mathematician) cannot be achieved. Among them is the theorem of Tarski stating the undefinability of the concept of truth and the theorem of Löwenheim and Skolem showing that a mathematical structure cannot be adequately and uniquely described by a formalized theory (a theory having a model has in fact many various models). Tarski wrote (1969, p. 74):

> It was undoubtedly a great achievement of modern logic to have replaced the old psychological notion of proof, which could hardly ever be made clear and precise, by a new simple notion of a purely formal character. But the triumph of the method carried with it the germ of a future setback.

Considerations concerning formal proofs enlightened also the role played by infinity in mathematics, in particular in the process of proving. Gödel's results show that finite/finitistic methods of syntactical formal provability do not exhaust the variety of mathematical truth. In fact, if one wants to obtain a complete theory then some infinite/infinitistic rules (such as the ω-rule) are necessary. Recall that the ω-rule is an inference rule with infinitely many premisses, i.e. it is the following rule:

$$\frac{\varphi(0), \varphi(1), \varphi(2), \ldots, \varphi(n), \ldots (n \in \mathbb{N})}{\forall x \varphi(x)}$$

In mathematical research practice, nobody restricts himself/herself to finite methods; on the contrary, any correct methods, among them infinite (in particular set-theoretical and semantical), are applied. Being not so ideal as it was hoped, formal proofs play an important role in metamathematics, i.e. in the study of mathematical theories or of mathematics as a collection of theories – but not only there.

They enable also the automatization and mechanization of proofs in mathematics, hence they make possible the construction of automated proofs and the verification of proofs by computers. Verification of (formal) proofs is possible because the relation "x is a proof of y" is – as was shown in mathematical logic – recursive, hence effective and can be implemented. On the other hand, constructing and finding proofs is not an effective (recursive) procedure; there is no universal method of doing this (as results of Turing, Church and Gödel show). In fact, it is only recursively enumerable. Hence every formal proof is a result of a creative invention of a human being. One can say that "[f]ormalization is about checking, and not about discovery" (cf. Wiedijk 2008, p. 1414).

Observe that the concept of a formalized proof is one for all theories; it is in a sense a uniform concept. It is independent of subjective, cultural and sociological elements and factors. Moreover, the completeness theorem of Gödel states that the logical means of the first-order logic (first-order predicate calculus) are sufficient.[27] This concept enables us to make the concept of a proof more objective. It also makes possible the precise study of provability in mathematics – under the assumption that the logical concept of a proof reflects all important and essential features of proofs from the research practice of mathematicians. One can prove results stating that a given statement is not a theorem of a given theory, i.e. that there exist no (formal) proof of a given statement or that a given sentence is (formally) undecidable (in a given theory). It enables also the study of important metamathematical properties of mathematical theories such as consistency, completeness, independence of axioms or axiomatizability in a given way, etc.

The concept of a formal proof is also helpful in the philosophy of mathematics. It can be used in attempts to answer the question about the existence and character of mathematical objects as well as in considerations concerning the epistemology of mathematics. On the other hand, all philosophies of mathematics reducing mathematics to formalized axiomatic theories (among them logicism and formalism) have a reductionist character and do not take into account the actual research practice of mathematicians. Their aim is to justify mathematics and not to explain the real mathematical practice. It is worth noting that in recent trends in the philosophy of mathematics still more and more attention is paid to the study of the research practice in mathematics – one takes into account various sociological, psychological and cultural factors. Unfortunately, it is done only by the analysis of particular single discoveries and achievements, hence by case studies. There are no general conceptions. But is it possible to develop such general conceptions?

---

27 Note that in the case of, e.g., second-order logic the situation is different – one does not have here the completeness phenomenon.

In fact, formal proofs are connected rather with foundational studies than with research practice. Observe that a formal proof does not give an understanding; it does not explain the deep reasons of a theorem. They are also not suitable for the practice – they are simply too long, they are too tedious and painstaking. In such a proof, the underlying intuition may get lost. Formalized mathematics may be also more error-prone than the usual informal one – in fact, formal manipulations may become very complicated. As Bourbaki (1968) wrote:

> If formalized mathematics were as simple as the game of chess, then once our chosen formalized language had been described there would remain only the task of writing out proofs in this langauge, [...]. But the matter is far from being as simple as that, and no great experience is necessary to perceive that such a project is absolutely unrealizable: the tiniest proof at the beginning of the Theory of Sets would already require several hundreds of signs for its complete formalization. [...] formalized mathematics cannot in practice be written down in full, [...]. We shall therefore very quickly abandon formalized mathematics, [...].

CadwalladerOlsker (2011, p. 42) writes:

> a purely formal proof [...] cannot be very complex without becoming so lengthy as to be incomprehensible to a human reader. Such a formal proof is rarely able to be explanatory, and may only be convincing to the degree that it can be read and understood by the reader or checked by a computer.

Add that a transcription of a single traditional (hence informal) proof into a formal one is a major undertaking.

Conclusion

We have shown that one has two concepts of a proof in mathematics: an informal one used by mathematicians in their usual research practice and the concept of a formal or formalized proof used mainly in logic and the foundations of mathematics. The first one is not defined precisely; it is simply practised and any attempts to define it fail. It is – so to speak – a practical notion. It has a psychological, sociological and cultural character. The second one is precisely defined in terms of logical concepts. Hence, it is a logical concept having rather theoretical than practical character. The first one has – in a part at least – semantical character, and the second is entirely syntactical in nature.

Hence this situation can be compared with the situation concerning Church-Turing thesis. This thesis states the equivalence of two concepts: effective computability (in the intuitive sense) and recursiveness (or Turing computability or computability in the sense of Markov or any other precisely defined and equivalent sense). As is known, this equivalence cannot be proved with the degree of precision usual and required in mathematics. The reason is that one part of this equivalence contains an intuitive vague concept formulated in the everyday language and the other a precise concept defined in the language of mathematics (cf. Murawski 2004b as well as Murawski and Wolenski 2006).

With such a situation, one has to do also in other parts of mathematics – see, for example, the concepts of function, of truth, of logical validity or of limit (cf. Mendelson 1990). In fact, till the 19th century a function was tied to a rule for calculating it, generally by means of a formula. In 19th and 20th centuries mathematicians started to define a function as a set of ordered pairs satisfying appropriate conditions (hence a function is identified here with its graph). The identification of those notions, i.e., of an intuitive notion and the precise set-theoretical one, can be called "Peano thesis". Similarly, "Tarki's thesis" is the thesis identifying the intuitive notion of truth and the precise notion of truth given by Tarski in (1933). The intuitive notion of a limit widely used in mathematical analysis in the 18th century and then in the 19th century applied by A. Cauchy to define basic notions of the calculus has been given a precise form only by K. Weierstrass in the language of "$\varepsilon - \delta$". There are many other such examples: the notion of a measure as an explication of area and volume, the definition of dimension in topology, the definition of velocity as a derivative, etc.

Comparing the two concepts of a proof in mathematics one can formulate a thesis stating that they are equivalent – one can call such a thesis a proof-theoretical thesis.[28] As in the case of Church Thesis, no precise and strict proof of it can be given.[29] One can only formulate arguments in favour or against it. The main argument for this thesis is the conviction – being popular among mathematicians and logicians – that every "normal" mathematical proof can be formalized, i.e. can be written as a formal proof in a suitable axiomatic theory. There are, of course, no general rules describing how this can and should be done. In fact, a formalization of an informal proof requires often some original and not so obvious ideas.

In this way, we come to the following conclusion: there are two concepts of a proof in mathematics. They play different but complementary roles: formal proofs

---

28 Add that this formulation is consciously rather vague – e.g. it is not specified here in which formal theory (or theories) formal proofs should be constructed and what should be the underlying logic. A similar but stronger thesis was formulated by Barwise (1977, p. 41) under the name "Hilbert's Thesis" where he wrote: "... the informal notion of provable used in mathematics is made precise by the formal notion provable in first-order logic. Following the sug[g]estion of Martin Davis, we refer to this view as Hilbert's Thesis". This thesis says that first-order logic is the logic of mathematics.

29 Note that in the case of Church Thesis the formal framework is precisely specified: the intuitive notion of computability should be captured by one specific formal model of computation.

are used mainly in metamathematical and logical considerations, whereas informal proofs are used in the research practice of mathematicians.

Note. The paper is based on my talks at the conferences *Philosophy in Science* (Kraków, Poland, 2012) and *Proof* (International Conference within the frame of Humboldt-Kollegs, Bern, Switzerland, 2013). I would like to thank the referee for helpful comments and suggestions.

The Status of Church's Thesis

Co-authored by Jan Woleński

The Church's Thesis can be simply stated as the following equivalence:

(CT) A function is effectively computable if and only if it is partially recursive.[30]

Thus (CT) identifies the class of effectively computable or calculable (we will treat these two categories as equivalent) functions with the class of partially recursive functions. This means that every element belonging to the former class is also a member of the latter class and reversely. Clearly, (CT) generates an extensional co-extensiveness of effective computability and partial recursivity. Since we have no mathematical tasks, the exact definition of recursive functions and their properties is not relevant here. On the other hand, we want to stress the property of being effective computable, which plays a basic role in philosophical thinking about (CT).[31]

A useful notion in providing intuitions concerning effectiveness is that of an algorithm. It refers to a completely specified procedure for solving problems of a given type. Important here is that an algorithm does not require creativity, ingenuity or intuition (only the ability to recognize symbols is assumed) and that its application is prescribed in advance and does not depend upon any empirical or random factors. Moreover, this procedure is performable in a finite number of steps. Thus a function $f : \mathbb{N}^n \to \mathbb{N}$ is said to be effectively computable (briefly: computable) if and only if its values can be computed by an algorithm. Putting this in other words: a function $f : \mathbb{N}^n \to \mathbb{N}$ is computable if and only if there exists a mechanical method by which for any n-tuple $(a_1, \ldots, a_n)$ of arguments, the value $f(a_1, \ldots, a_n)$ can be calculated in a finite number of prescribed steps. Three facts should be stressed here: (a) no actual human computability or empirically feasible computability is assumed in (CT); (b) functions are treated extensionally,

---

30 We do not enter in the history of (CT) and its various formulations. Some people prefer to speak of the Church-Turing thesis: a function is effectively calculable if and only if it is Turing computable. In fact, Church used the concept of λ-definability. The principal historical details are to be found, e.g., in Gandy (1988), Schulz (1997, pp. 159–171) (this book provides an extensive analysis of Church's Thesis, including the problems discussed in this paper) and Murawski (2004b). We choose the formulation via recursive functions for its applications in logic.

31 Henceforth, "computable" is an abbreviation for "effectively computable" and "recursive" – for "partially recursive".

i.e. a function is identified with an appropriate set of ordered pairs; (c) the concept of computability has a modal parameter ("there exists a method", "a method is possible") as its inherent feature.

Typical comments about (CT) are as follows:

> (i) Church's thesis is not a mathematical theorem which can be proved or disproved in the exact mathematical sense, for it states the identity of two notions only one of which is mathematically defined while the other is used by mathematicians without exact definition.[32] (Kalmár 1959, p. 72)
>
> (ii) While we cannot prove Church's thesis, since its role is to delimit precisely a hitherto vaguely conceived totality, we require evidence that it cannot conflict with the intuitive notion which is supposed to be complete; i.e. we require evidence that every particular function which our intuitive notion would authenticate as effectively calculable is [...] recursive. The thesis may be considered a hypothesis about the intuitive notion of effective calculability; in the latter case, the evidence is required to give the theory based on the definition the intended significance. (Kleene 1952, pp. 318–319), Kleene 1967, p. 232)
>
> (iii) This is a thesis rather than a theorem, in as much as it proposes to identify a somewhat intuitive concept phrased in exact mathematical terms, and thus is not susceptible of proof. But very strong evidence was adduced by Church, and subsequently by others, in support of the thesis.
>
> It [(CT) – and other similar characterizations – our remark, R.M. and J.W.] must be accepted or rejected on grounds that are, in large part, empirical. [...]. Church's Thesis may be viewed as a proposal as well as a claim, that we agree henceforth to supply previously intuitive terms (e.g., "function computable by algorithm") with certain precise meaning. (Rogers 1967, p. 20)

These three quotations shed some light on several problems raised by (CT). Firstly, we can and should ask for evidence for it. We take the standard position that the implication from recursivity to computability (every recursive function is computable) is obvious and the opposite implication from computability to mathematical definition of effective calculability, i.e., recursivity (every computable function is recursive), has a sufficient justification.[33] Secondly, one can ask for the fate of (CT) in some logical framework, in particular, in intuitionistic or constructive systems (see Kleene 1952, pp. 318, 509–516, Kreisel 1970, McCarty 1987), but we

---

32 Kalmár argues against (CT), but we do not enter into this question.
33 The accessible evidence for (CT) is collected in every textbook of logic having a part on recursion theory or a monograph about the topic. See, e.g., Kleene (1952, pp. 317–323), Kleene (1967, pp. 232–242), Murawski (1999, pp. 85–87), Murawski (2004b) and Folina (2006). Since the implication (a) "if a function is recursive, then it is computable" does not raise doubts the Church thesis is sometimes reduced to (b) "if a function is computable, then it is recursive". On this route, (a) is called "the converse Church's thesis". We do not follow this custom.

entirely neglect this question.[34] Thirdly, there are various special problems, mostly philosophical, we believe, related to (CT). Does this thesis support mechanism in the philosophy of mind or not (see Webb 1980)? How is it related to structuralism in the philosophy of mathematics (see Shapiro 1983)? We also neglect this variety of questions, except eventual parenthetical remarks aimed at exemplification. Fourthly, and this is our main concern in this paper, there arises the problem of the status of (CT). We split this topic into two subproblems. (CT) can be considered from the point of view of its function in mathematical language or various conceptual schemes. The second subproblem focuses on the character of (CT) as a statement or sentence. To be more specific, we note that one of the views considers (CT) as a definition. This gives an illustration of the former subproblem. However, independently whether (CT) has the status of a definition or not, it is captured by a sentence. Now, we can ask whether this sentence is analytic or synthetic, *a priori* or *a posteriori*. This provides an illustration of the latter subproblem. Although both subproblems are closely related, their separation, even relative, makes the analysis of the status of (CT) easier.

The following views about the function of (CT) can be distinguished :[35]

(A) (CT) is an empirical hypothesis,
(B) (CT) is an axiom or theorem,
(C) (CT) is a definition and
(D) (CT) is an explication.

Ad. (A) (CT) can be considered as referring to human (possibly idealized) abilities. Hence Church's Thesis is connected with questions concerning relations between mathematics and material or psychic reality. The last view is represented by Post, who regarded the notion of computability as a notion of a psychological nature (Post 1965, pp. 408, 419; page-reference to the first edition):

> [...] for full generality a complete analysis would have to be made of all the possible ways in which the human mind could set up finite processes for generating sequences. [...] we have to do with a certain activity of the human mind as situated in the universe. As activity, this logico-mathematical process has certain temporal properties; as situated in the universe it has certain spatial properties.

Post maintained that (CT) should be interpreted as a natural law and insisted on its empirical confirmation (similarly DeLong 1970, p. 195). However, one should be very careful with such claims. In fact, we have no general and commonly accepted psychological theory of human mental activities, even as far as the matter concerns

---

34 Note, however, that although (CT) defines the concept of effective computability, it functions within classical, i.e. not constructive metalogic.
35 We order these proposals according to their plausibility from our point of view.

the scope of computation. Hence, it is very difficult to think about (CT) as an empirical hypothesis about the human mind and its abilities. Eventually (CT) can be considered as related to mechanism in the philosophy of mind (see above) and even as confirming this view. Although we do not deny that (CT) has an application in philosophy, we do not think that this thesis as a tool for a philosophical analysis of the mind body problem functions as a genuine empirical hypothesis. We think that the use of (CT) for supporting mechanism in the philosophy of mind is very similar to deriving (or not) indeterminism from the principle of indeterminacy. We have to abstain from further remarks concerning this interpretation of (CT), because it would require a longer metaphilosophical discussion.

Ad. (B) One should distinguish two understandings of axioms or theorems. Firstly, axioms can be considered as principal metatheoretical assumptions. When Kreisel (1970) considers (CT) as a kind of reducibility axiom for constructive mathematics, he uses the concept of axiom in this sense. Yet, we are inclined to think that his comparison of (CT) with "hypothesis" of $V = L$ in set theory is rather misleading. In the case of $V = L$, there is no need to use quotes and say "hypothesis", because we have to do with a set-theoretical axiom, whereas Kreisel's analysis does not result in an axiomatic system of set theory. This leads to the second use of the term "axiom", referring to an element of an axiomatic system. A similar problem arises with respect to the category of being a theorem. Mendelson (1990, p. 230) writes:

> I would like to challenge the standard interpretation of CT as an unprovable thesis. My viewpoint can be brought out clearly by arguing that CT is another in a long line of well-accepted mathematical and logical "theses", and that CT may be just as deserving of acceptance as those other theses. Of course, these theses are not ordinarily called "theses" and that is just my point.

Further, Mendelson mentions the following examples as comparable with (CT): (a) the set-theoretical definition of function as a kind of relation; (b) Tarski's definition of truth; (c) the definition of logical validity and (d) Weierstrass's definition of limit. And Mendelson continues (1990, p. 232):

> [...] it is completely unwarranted to say that CT is unprovable just because it states an equivalence between a vague, imprecise notion (effectively computable function) and a precise mathematical notion (partial-recursively function).

Mendelson gives three more specific arguments as supporting his view (1990, pp. 232–233):

> The concepts and assumptions that support the notion of partial-recursive function are, in an essential way, no less vague and imprecise than the notion of effectively computable function; the former are just more familiar and are the part of a respectable theory with

connections to other parts of logic and mathematics. (The notion of effectively computable function could have been incorporated into an axiomatic presentation of classical mathematics, but the acceptance of CT made this unnecessary). [...].

The assumption that a proof connecting intuitive precise mathematical notions is impossible is patently false. In fact, half of (CT) (the "easier" half), the assertion that all partial recursive functions are effectively computable, is acknowledged to be obvious in all textbooks in recursion theory. A straightforward argument can be given for it. (The so-called initial functions are clearly effectively computable [...]. Moreover, the operations of substitution and recursion and the least-number operator lead from effectively computable functions to effectively computable functions. [...].) This simple argument is as clear as a proof as I have seen in mathematics, and it is proof in spite of the fact that it involves the intuitive notion of effective computability. [...].

Another difficulty with the usual viewpoint concerning CT is that it assumes that the only way to ascertain the truth of the equivalence asserted in CT is to prove it. In mathematics and logic, proof is not the only way in which a statement comes to be accepted as true. Of course, this is a consequence of the truism that not all truths can be proved; proofs must assume certain axioms and rules of inference.

Of course, it is true that in mathematics there occur not only formal proofs but also other methods of justification accepted by mathematicians. Thus, Mendelson is right provided that the concepts of axiom and theorem are taken liberally.

However, Mendelson's view raises essential doubts (see also Shapiro 1993, Folina 2006; Mendelson 2006 seems to take a more careful position). First of all, any discussion of (CT) and similar theses in the framework of logic or the foundations of mathematics requires an appeal to the metamathematical notions of proof, axiom and theorem; because, in the opposite case, the development of logic in any textbook of mathematical logic, including celebrated Mendelson (1970), would be redundant. To prove (in a formal way) Church's Thesis one should construct a formal system in which the concept of computability would be among the primitive notions and which would be based (among other things) on axioms characterizing this notion. The task would be then to show that computability characterized in such a way coincides with recursiveness. But another problem would appear now: namely the problem of showing that the adopted axioms for computability do in fact reflect exactly the properties of computability (intuitively understood). Hence, we would arrive at Church's Thesis again, though at another level and in another context. It should be also noted that Kleene and Rogers (see above) spoke about theorems in another sense, namely, as sentences (formulas) provable in an explicit formal (or formalizable) axiomatic system. "Being provable" means here being derivable by explicit inferential devices specified within an assumed system of logic. This treatment enables one to include axioms as a kind of theorems: logical axioms and their consequences are provable from the empty class of sentences, but extralogical axioms are their own proofs; that is, if A is an extralogical axiom of a system **S**, its proof in **S** consists solely of A itself (the same can be said about logical axioms as well).[36] Secondly, even if the counterparts of intuitive and informal notions occur in axioms or theorems, they lose, entirely or partially, their ordinary or colloquial meanings (senses) and begin to function as axioms (postulates) and definitions of a given **S**. If a chemist defines water as H_2O, the substance denoted by "water" in theoretical chemistry is just H_2O and not water in the ordinary sense. Similarly, sets in axiomatic set theory are understood according to axioms, but not as collections in colloquial language or even in informal mathematics. Moreover, we find the phrase "concepts and assumptions that support the notion" very unclear. The circumstances supporting notions are always of a psychological nature, but, when we go to theories, concepts are supported only by the axioms in which they occur. Thirdly, we do not agree that the concept of recursive function is more familiar than the concept of computable function. Fourthly, the fact that the former is an element of a respectable mathematical theory, but the latter is not, constitutes just the point. Fifthly, although "the notion of effectively computable function could have been incorporated into an axiomatic presentation of classical mathematics", we doubt whether another way than the repetition of the theory of recursive functions (or an equivalent theory) appears as proper here. Thus, we think we can conclude that Mendelson's position stems from the various uses of the concepts of proof, axiom and theorem.

Yet we cannot ignore the role of (CT) in proofs. Rogers (1967, p. 21) speaks about "proofs by Church's Thesis" as relying on informal methods and, in this case, all evidence supporting (CT).[37] In fact, these proofs translate results achieved in the theory of recursive functions into that about computability. In particular, (CT) is suited for establishing negative results in this respect. When one shows that a given function is not recursive, then one can conclude that it cannot be computed. Thus, we prove that the set of theorems of predicate calculus is not decidable because it is not recursive, where decidability is understood as a kind of computability. Perhaps the case of the first incompleteness theorem is the most interesting. Kleene (1987, p. 495, note 4) (see also Krajewski 2006) observes that if **S** is an ω-consistent and complete system of formal number theory, then (CT) does not hold. Hence, we obtain, even by intuitionistic logic, i.e. fully constructively, that if it is not true that ω-consistency of **S** implies its incompleteness, then (CT) does not hold.

---

36 There are also theorists who regard Church's Thesis as proved. Gandy (1988) argues that Turing's direct argument pointing out that every algorithm can be simulated on a Turing machine proves a theorem. He regards this analysis to be as convincing as typical mathematical work. This concerns (CT) in terms of Turing machines.

37 This function of (CT) was indicated by Church in (1936a): "[...] the author has proposed a definition of the commonly used term 'effectively calculable' and has shown on the basis of this definition that the general case of the *Entscheidungsproblem* is unsolvable".

Webb (1980, p. 208) considers the connection between the incompleteness of arithmetic and (CT) as very deep (emphasis follows the original; the letter **S** is introduced instead of *F*):

> [...] true but unprovable sentences are *just the guardian angels which look after the formalization of effectiveness found in any suitable* **S**; they protect (CT) from refutation by the diagonal argument.

The irrefutability of (CT) by diagonalization means that no nonrecursive function can be defined in **S**, provided that it is incomplete.

Kleene (1952, p. 302; 1967, pp. 250–254) proved the so-called generalized Gödel's theorem:

(**GGT**) There is no correct and complete formal system for the predicate "being a computable proof in **S**".

This does not support, however, that (CT) functions as an axiom or theorem. The proof of (GGT) is by Church's Thesis in the Rogers sense (see above) and, as Kleene himself (1967, p. 252) points out, can be easily converted into a derivation without appealing to (CT). Moreover, the actual significance of the protection of (CT) by the incompleteness phenomenon should be properly understood. We can make **S** complete by adding the ω-rule, which is non-finitary. This immediately leads to a new proof-predicate, which is not computable in the sense of (CT), i.e. recursive. Thus **S'** = **S** + the ω-rule generates a wider class of number–theoretical functions than **S** itself. Now we can generalize (CT) in order to capture new functions; this would lead to a different concept of recursivity. We have the following version of the incompleteness theorem (see Smullyan 1993, p. 41; we assume that **S** is consistent):

(**GT'**) If **S** is an axiomatizable system in which some non-recursive set is representable, then **S** is incomplete.

By contraposition, we obtain:

(**GT"**) If **S** is complete, then **S** is non-axiomatizable or no nonrecursive set is representable.

Since **S'** is axiomatizable and complete, it has to represent some non-recursive set, where "recursive" means what is usually accepted by this adjective. Hence, **S'** extends the class of computable functions beyond (CT). The connection of incompleteness and (CT) is actually very deep but not purely deductive. In particular, it appears that the assumptions of Gödel's theorem and (CT) are exactly the same as far as the matter concerns the finitary character of deductive devices. Kleene (1987, p. 495, note 4) rightly points out that the dependence "if it is not true that ω-consistency of **S** implies its incompleteness, then (CT) does not hold" asserts (we prefer to say "suggests") the absurdity of the completeness of **S**, "rather than giving a specimen of an undecidable formula". Putting this in another words, the completeness of **S** would abolish its character as a system based on recursive (= computable) machinery.[38] However, assume that one started with **S'** as the proper number theory, arguing that mathematicians use the ω-rule, but keeping the standard intuition about effectively calculable functions. Obviously, one could say "well, I understand calculable functions as corresponding to recursive functions associated with **S** = **S'** – the ω-rule". The effect is exactly the same as in the case of (CT).

Ad. (C) Church (1936, pp. 90, 100; the italics follows the original; page-reference to the reprint) introduced (CT) in the following way:

> The purpose of the present paper is to propose a definition of effective calculability [...].
> We now define the notion of [...] *an effectively calculable function of positive integers by identifying it with the notion of a recursive function of positive integers*.

A similar treatment was suggested by Gödel (1946, p. 150; page-reference to the reprint):

> [...] one has for the first time succeeded in giving an absolute definition of an interesting epistemological notion, i.e., one not depending on the formalism chosen.

If we look at the formulation of (CT), its structure does not exclude considering it as a definition. To see this we can convert (CT) into (CT'): a calculable function is a function which is recursive.

Now "calculable function" appears here as the definiendum, but the phrase "a function which is recursive" serves as the definiens. However, closer inspection immediately leads to some questions. The word "function" seems to have exactly the same meaning in both parts of (CT'). Hence, we cannot say that the concept of function is a genus which is determined by a *differentiam specificam*. This means that (CT') does not fall under the classical formula that *definition fit per genus proximum et differentiam specificam*. It appears rather that (CT) offers a definition of calculable by the category of recursive, independently of whether it is treated intensionally or extensionally.

Since the classical formula of framing all definitions by the proximate genus and the specific difference proves to be inadequate, it is nothing wrong with the fact that (CT) lacks of this character. More important worries concern how (CT) should be

---

38 There is also another way (see Kleene 1967, p. 252, although our argument goes further) to justify the same conclusion. Instead of adding the ω-rule, we could take all arithmetical truths in the standard model as axioms. Although the resulting system is complete and has a finite deductive machinery for proving every axiom by itself (otherwise speaking: every axiom has a proof consisting of the single element, i.e. the axiom in question), it is based on a non-recursive axiomatic base. Hence, it defines a class of functions, which is not recursive in the usual sense.

qualified as a definition. Let us recall some traditional distinctions. Firstly, we distinguish nominal and real definitions, i.e. those of words and things, respectively. Secondly, there are analytic (reporting), regulative and synthetic (designer) definitions. Now it is difficult to locate (CT) under these rubrics. Certainly, it is not a nominal definition, because nobody regards its acceptance or rejection as a matter of taste, convenience, etc. Even if one says that (CT) concerns the use of "calculable" to some extent, its obvious actual enterprise consists in providing an objectual (substantial) characterization. But what does "objectual" mean in this context? To touch the essence of a calculable function? Is the essence of calculable the same as the essence of recursive? We claim that since no straightforward answer can be given to these questions, any reasonable decision whether (CT) considered as a definition is nominal or real is hopeless. We qualify in a similar manner the solution of the question whether (CT) is analytic, regulative or synthetic. To start with the middle case, "calculable" is not a vague adjective as "bald" or "short", because functions are or are not calculable, without being such to some degree, which should be made precise.[39] Since "calculable" needs to be defined in order to put it into a formal mathematical theory, the alleged definition cannot be analytic, but since (CT) is certainly not an arbitrary claim, this prevents its treatment as a synthetic proposal.

There is still another reason to doubt whether (CT) is a definition in the proper sense. We do not agree that the expression "calculable function" occurs only in colloquial language, understood as the speech of ordinary people; eventually, "colloquial" means here "functioning in the language of ordinary mathematics" (we will return to this question). Now mathematical definitions can similarly be considered as axioms, proofs or theorems (see above). In one sense, a definition can explain some more or less established intuitions. For example, one can say that to obtain a natural number *n*, one should start with 0 and subsequently apply the operation of adding 1 by performing *n* times this step. This definition of the natural number n is to be then formalized by the Peano axioms. There is no other way to check the correctness of such steps than by appealing to a conformity between informal sources and formalizations. It is perhaps interesting that an informal definition of a natural number is not subjected to an evaluation as correct or not, although it is very far from being arbitrary. On the contrary, it is determined by quite definite intuitions stemming from ordinary counting. However, if we add definitions of addition and multiplication to formal (or formalizable) number theory, it is easy to establish whether they are correct or not. Analogically, (CT), as far as it functions in the language of informal mathematics, can be more or less supported by various

---

39 We do not exclude the situation that some functions are approximately calculable, but admitting that possibility would change the meaning of calculability.

data, but we are not able to establish its correctness in any other way than by simultaneous acting with its informal as well as formal exposition. What we can do with full precision is to assert that the definition of recursive function is correct or not, but we do this within formal (formalizable) mathematics. Our conclusion is that (CT) as a definition functions only as an exposition of intuitions and thereby is not a definition in the proper sense.

Ad. (D) The concept of rational reconstruction is closely related to the notion of explication introduced by Carnap in the following way (Carnap 1952, pp. 3–8; the quotation is taken from p. 3 and p. 7; the emphases follow the original):[40]

> The task of explication consists in transforming a given more or less inexact concept into an exact one or, rather, in replacing the first by the second. We call the given concept (or the term used for it) the *explicandum*, and the exact concept proposed to take the place of the first (or the term proposed for it) the *explicatum*. The explication belongs to everyday language or to a previous stage in the development of scientific language. The explication must be given by explicit rules for its use, for example, by a definition which incorporates it into a well-constructed system of scientific either logimathematical or empirical concepts. [...] the task of explication may be characterized as follows. If a concept is given as explicandum, the task consists in finding another concept as its explicatum which fulfils the following requirements to a sufficient degree:
>
> 1. The explicatum is to be *similar to the explicandum* in such a way that, in most cases in which the explicandum has so far been used, the explicatum can be used; however, close similarity is not required, and considerable differences are permitted.
> 2. The characterization of the explicatum, that is, rules for its use (for instance, in the form of a definition), is to be given in an *exact* form, so as to introduce the explicatum into a well-connected system of scientific concepts.
> 3. The explicatum is to be a *fruitful* concept, that is, useful for the formulation of many universal statements (empirical laws in the case of a nonlogical concept, logical theorems in the case of a logical concept).
> 4. The explicatum should be as *simple* as possible; this means as simple as the more important requirements (1), (2), and (3) permit.[41]

Since we can admit that the idea of calculable function needs to be explained, (CT) conforms to the task of explication in Carnap's sense, and we can take the concept of calculable function as the ex- plicandum, and the notion of recursive function as the proposed explicatum for the former. Clearly, it leads to replace an intuitive and not quite exact concept by a fully legitimate mathematical category.

---

40 We prefer to speak about explication than rational reconstruction. The second concept is used by Mendelson (1990, p. 229) and Schulz (1997, p. 182).
41 Carnap gives i.a. the following examples of this procedure: the explication of "warm" (explicandum) by "having such and such temperature" (explicatum) and "logical probability" (explicandum) by "the degree of confirmation" (explicatum).

Doubtless, the requirements 2. and 3. are satisfied, because the rules for the explicatum are given in an exact form; and it expresses a concept which is useful for the formulation of many universal theorems. Since the explicatum replaces the explicandum in all known applications, the most important part of 1. is fulfilled. This similarity of the explicatum to the explicandum and the simplicity of the former might be problematic, but this question does not need to worry users of (CT). A clear advantage of the method of explication consists in its neutrality with respect to the nominal/real distinction. Independently of whether we take the concept or the term as an explicandum (explicatum) the result is the same. In our case, it means that (CT) explicates the notion of calculable function as well as the term "calculable function".

Although treating (CT) as an explication appears to us essentially better than (A)–(C), we consider it to be somewhat unsatisfactory. Our reservations do not concern the clear vagueness of 1. and 4. in Carnap's formulation. In fact, one should postulate similarity and simplicity in question, although no exact measure is possible here. In order to formulate our position let us return to the use of "calculable function". As we already noted, we do not agree that it belongs to the vocabulary of a colloquial speech. However, there is more to say. As we observed at the beginning of this paper, the concept of computability has a modal parameter, expressed by phrases, like "there exists a method" or "a method is possible". Although such notions always raise doubts concerning their content, boundaries, etc., it would be incorrect to say that they have no standard meaning, particularly if they belong to specialized languages. This matter can be explained by using Ryle's analysis of the adjective "ordinary" (see Ryle 1953). He distinguishes two expressions, namely (a) the use of ordinary language; (b) the ordinary use of expressions, and observes that "ordinary" has a different meaning in (a) and (b). He writes (pp. 302–304):

> When people speak of the use of ordinary language, the word "ordinary" is in implicit or explicit contrast with 'out-of-the-way', 'esoteric', 'technical', 'poetical', 'notational' or, sometimes, 'archaic'. 'Ordinary' means 'common', 'current', 'colloquial', 'vernacular', 'natural', 'prosaic', 'non-notational', 'on the tongue of Everyman', and usually in contrast with dictions which only a few people know how to use, such as the technical terms, or artificial symbolisms of lawyers, theologians, economists, philosophers, cartographers, mathematicians, symbolic logicians and players of Royal Tennis. There is no sharp boundary between 'common' and 'uncommon', 'technical' and 'untechnical' or 'old-fashioned' and 'current'. [...] But in other phrase, 'the ordinary use of the expression "...", 'ordinary' is not in contrast with 'esoteric', 'archaic' or 'specialist', etc. It is in contrast with 'non-stock' or 'standard'. [...] If a term is a highly technical term, most people will not know its stock use or, *a fortiori*, any non-stock use of it either, if it has any. [...] A philosopher who maintained that certain philosophical questions are questions about the ordinary or stock uses of certain expressions would not therefore be committing himself to the view that they are questions about the uses of ordinary or colloquial expressions. He could admit that the noun 'infinitesimals' is not on the lips of Everyman and still maintain that Berkeley was examining the ordinary

or stock use of 'infinitesimals', namely the standard way, if not the only way, in which this word was employed by mathematical specialists. Berkeley was not examining the use of a colloquial word; he was examining the regular or standard use of a relatively esoteric word. We are not contradicting ourselves if we say that he was examining the ordinary use of an unordinary expression.[42]

It is fairly problematic how to draw a borderline between the functioning of "computable" or "calculable" as common adjectives and situations in which they belong to a very specialized mathematical vocabulary. On the other hand, Church examined the standard (stock) use of the term "computable function" in the language of informal mathematics.[43] However, this does not mean that the ordinary (standard) use suffices for some special tasks. If this becomes the case, the standard use of an expression requires a move toward making it more precise and its explication appears as profitable or even necessary in some circumstances.[44] We think that explications of the ordinary (standard) use of expressions consists in their normalization by exact conceptual means drawn from well-established theories. Normalizations in this sense usually cohere with definite intuitions and this protects them from being arbitrary, but the conditions for their correctness remain partially open. Now it should be clear why Carnap's formulations of constraints 1.–4. employing phrases like "close similarity" or "as simple as possible" are subjected to various and mutually conflicting interpretations. Although Carnap himself was very optimistic about the effectivity of the method of explication, a modest position appears to be more justified. In general intuitive properties, like computability, cannot exactly coincide with precise ones, like recursiveness or its formal equivalents. And this must be always remembered when we adopt Carnapian requirements as proper for the practice of normalization.[45] In the case of

---

42 To prevent possible misunderstandings, we note that we do not subscribe to Ryle's metaphilosophy on which most philosophical questions concern the ordinary (standard) use of expressions belonging to ordinary (common) language.
43 In fact, all of Mendelson's examples (see above), perhaps except Tarski's truth-definition, are of the same character.
44 The reasons for that cannot be given in advance. Avoiding paradoxes, building theories, needs to be more precise, etc. belong to typical causes. Although predictions are difficult here, scientific (or even colloquial) practice has decisive significance. However, we do not agree, even very strongly, with the following opinion (Kalmár 1959, p. 79): "There are pre-mathematical concepts which must remain pre-mathematical ones, for they cannot permit any restriction imposed by an exact mathematical definition.Among these belong, I am convinced, such concepts as that of effective calculability, or of solvability [...]". On the contrary, we are convinced that 'must' is entirely unjustified in this context.
45 Just this character of normalizations causes typical conditions of the correctness of definitions not to apply to the former. Although we can speak about definitions in a wider sense, which also covers normalizations, the differences here are essential.

(CT), we should also note that it replaces a modal property of functions by one that is non-modal and is definable in arithmetic. This feature of (CT) appeals to an intuition that Peano arithmetic and set theory accurately reflect the possibilities of computation. We are inclined to consider eliminating modalities for purely extensional concepts as a very typical aspect of normalizations.[46] Understanding (CT) as a result of normalization explains why it is sometimes considered as problematic and why there is no hope for changing this situation.[47]

We now pass to the second issue concerning the status of (CT), namely its character as a sentence. The choice of the position from the variety (A)–(D) determines the direction of analysis of how to solve the second subproblem to some extent. If one selects the position (A) as proper, (CT) must be understood as a genuine empirical statement, being a subject of confirmation or disconfirmation by empirical data as any other sentence of this type. The choice of (B) qualifies (CT) as a formula expressing a mathematical theorem and the further course depends on how such entities are conceived (analytic, tautologous, synthetic *a priori*, empirical, etc.). Similarly, (C) is associated with a view concerning the status of definitions. The matter looks differently for various kinds of definitions: reported, designer, regulative, real or nominal. Since we adopted (D) as a legitimate position, we only very roughly outline possible ways related to (A)–(C). As far as the matter concerns (D), we will defend the view that (CT) is an analytic sentence of a sort, as well as *a priori*.

The proposed solution requires a closer treatment of analyticity and apriority (we follow Wolenski 2004, although we restrict our considerations to mathematics). We begin with the former category. The crucial conceptual point consists in introducing two distinctions directed to the concept of analyticity. Firstly, we distinguish analytic sentences in an absolute and a relative sense, and both of these in a semantic and a syntactic understanding. Absolute analytic sentences in the semantic are those which are derivable from the empty set of sentences. By the completeness theorem, they can be identified with theorems (= validities) of first-order logic, i.e. formulas true in all models. The absolute analytic sentences in syntactic sense are those which are absolute in the semantic sense and belong to a set for which the decision problem has a positive solution (this category is not essential here). Having logic (= the set of first-order logical truths) well-defined, we say that A is relatively semantic analytic in a theory T if and only if it is true in all models

---

46 This view is completely independent of well-known arguments pro and contra modal scepticism, inspired by Quine. See Shapiro (1993) for a discussion of this matter in the context of (CT).

47 Although we deal with normalizations in science, we can find examples of them in simple situations taken from the Everyman perspective. In fact, every regulative definition, e.g. "being an adult", consists in a normalization.

of T; *A* is relatively syntactic analytic in T if and only if it is relatively semantic analytic in T and belongs to a decidable subset of T. Now we add analytic sentences in the pragmatic sense. Assume that T has standard as well as non-standard models. We say that *A* is an analytic sentence in the pragmatic sense in a theory T if and only if it is true in standard (intended) models of T. The label "in the pragmatic sense" stresses the fact that there are no purely general semantic criteria of standardness. If we qualify some models as standard, we appeal to certain intuitions; the adjective "intended" well displays this situation. Every analytic sentence in the pragmatic sense is relative, but we have inclusions inside the class of analyticities: absolute syntactic sentences are a subset of absolute semantic sentences; absolute semantic sentences are a subset of relative sentences, for any theory T (logic belongs to every set of sentences closed by logical consequence); absolute analytic sentences are a subset of analyticities in the pragmatic sense. Truths associated with the standard/non-standard distinction appear as perhaps the clearest kind of analytic sentences in the pragmatics sense. However, we have also another source of pragmatic analyticity, namely definitions. If a definition functions in a theory, its role as a generator of analytic sentences in the pragmatic sense is obvious. Observe that every definition occurs in some context, e.g. legal or even colloquial. If we define "being adult" as "being a person who has finished 18 years", we introduce a pragmatic analytic sentence. The qualification "pragmatic" is justified for being related to an intuition, for instance, that an adult should be conscious of his or her actions and their results, but this expectation is fulfilled by the average person who has finished 18 years.

Explications and normalizations in the above sense can be considered as a further source of pragmatic analyticity. This immediately leads us to (CT) and its status. Since (CT) occurs in the language of informal mathematics (more strictly: metamathematics), its context is fairly definite. We can say that (CT) functions in a conceptual scheme of mathematics and equates two concepts within this body. One notion, namely that of computable (calculable) function, is informal and intuitive, but the second, namely that of recursive function, is a component of the relatively semantic analyticities of arithmetic. Now (CT) as an equivalence normalizes the use (or meaning) of "computable function" in informal metamathematics by equating it with the use of "recursive functions". This move is governed by intuitions concerning the idea of computability and related to the identification in question. Although the resulting sentence, namely (CT), is not a tautology, it can be considered as an analytic sentence in the pragmatic sense. Such sentences are always surrounded by some empirical data, which motivate or strengthen the intuitions. In the case of (CT), the equivalence of many and various formal normalizations of the concept of computable function controls our initial intuitions as sufficiently correct, but it does not exclude that they could be revised. However, until a given normalization is accepted, (CT) should be considered as unconditionally true. We stress this point because there are views denying (CT) as a sentence with a definite truth-value (see Shapiro 1993 for a critical discussion of this position). Remembering that A is always true (false) in a language (better: in a theory or at least in a conceptual scheme) should be enough to admit that (CT) is true, if accepted, even as a pragmatic analyticity.

Is (CT) true *a priori* or *a posteriori*? The answer depends on how *a priori* (*a posteriori*) is understood. In the Kantian understanding, i.e. when *a priori* truths are entirely independent of experience, pragmatic analyticities are synthetic *a posteriori*. However, this theory of the *a priori* is too absolute for our tasks. We need a more liberal view, just modelled on the use of "*a priori*" in probability theory. Recall that an *a priori* probability is one which is ascribed to an event before studying this event in order to establish whether and how it is probable in the light of collected data. *A priori* probability can be established relatively to previously accumulated experience, for instance, when one assumes that the distribution of a statistical property in a population **P** is normal, although no investigation of **P** was performed. This approach suggests a distinction between the absolute and the relative *a priori*. Without deciding whether the absolute *a priori* occurs (we are inclined to think that only logic can serve as an example here), we admit that analytic sentences in the pragmatic sense are at the same time relative apriorities. Looking at normalizations from this perspective suggests that they also are simultaneously processes of apriorization. In fact, every conceptual system requires some elements accepted *a priori* for its cognitive stabilization. Consider (CT) once again. Mathematicians have various intuitions associated with computability and want to use them in a precise way. However, the concept of computable function is not enough in this respect; for example, it does not suffice to obtain results about decidability. Now formulating these results solely with the apparatus of recursive function theory does not suit mathematical intuitions (we do not evaluate, we only note the situation as we see it). Thus, mathematicians need a principle, a rule, etc., which would improve the situation. Even if it is formulated after collecting empirical data, the result consists in apriorization leading to the acceptance of (CT) as working before, i.e. just relatively *a priori*, investigations concerning a definitely stated problem, for example, the decidability of predicate calculus. We stress that the presence of empirical data is not at odds with analyticity and apriority, provided that both are understood as relative.[48]

Note. The financial support (for R. Murawski) of the Committee for Scientific Researches (Grant no. 1 H01A04227) is acknowledged.

---

48 We claim that a similar analysis concerns other famous proposals in meta-mathematics, e.g. Tarski's definition of truth and logical consequence.

Between Theology and Mathematics. Nicholas of Cusa's Philosophy of Mathematics

The aim of this chapter is to indicate the influence of theological and philosophical ideas on the philosophy of mathematics of Nicholas of Cusa (1401–1464). He was a mathematician but first of all a theologian. In fact the connections between theology and philosophy on the one side and mathematics on the other were in his case bilateral. He used mathematical language in explaining theological ideas and *vice versa* – some ideas and concepts coming from theology and philosophy were used by him to express his conceptions concerning philosophical questions and problems of mathematics. In this chapter we shall concentrate only on the second issue and try to show how some theological ideas were used by him to answer fundamental questions in the philosophy of mathematics.

Before we consider Nicholas' philosophy of mathematics, let us say some words about his life and activity.

He was born as Nicholas Kryffs or Krebs in Kues, now Bernkastel- Kues, about 30 km from Trier, an old town in the Palatinate, founded already by the Romans. Following the usual practice in a Latin speaking church environment, his name often appears as Nikolaus Cusanus, from the Latin name of the town. He was sent to Deventer, in the Netherlands where he probably attended a school of Brothers of Common Life, a Roman Catholic religious community founded in the 14th century. They influenced him with a mixture of mysticism and reason. In 1416 Nicholas matriculated at the University of Heidelberg where he studied liberal arts, particularly philosophy. The following year he went to the University of Padua where he studied canon law. In Padua he became a friend of Paolo dal Pozzo Toscanelli, who later became an important mathematician and astronomer. They remained friends throughout Nicholas' life. Thanks to his contacts with Toscanelli, Nicholas learned in Padua about the latest developments in mathematics and astronomy. He graduated with a doctorate in canon law from Padua in 1423. In 1425 he matriculated at the University of Cologne to study philosophy and theology. There he was introduced to the ideas of Pseudo-Dionysius, Albertus Magnus and Ramon Llull. After finishing his studies, he began his legal activity. In 1431–1437 he took part in the Council of Basel. In 1433 he wrote *De concordantia catholica* arguing that the Council's authority took precedence over that of the pope. In 1436 Nicholas changed sides, taking the pope's side. In 1438 pope Eugenius IV sent him as a member of a three-man delegation to Constantinople. Their aim was to set up a process leading to the eastern and western Churches reuniting. His activity led to temporary success. The stay in Constantinople was important for Nicholas also from the point of view of his scientific activity – he discovered there some important Greek manuscripts.

Between 1438 and 1448 Nicholas took part in several missions to Germany as papal envoy. Sometime between 1436 and 1440 he ordained and was named cardinal by pope Eugenius IV in 1446 in recognition of his work as papal envoy. The death of Eugenius IV caused that Nicholas had to wait till 1448 when pope Nicholas V made him a cardinal. He became the bishop of Brixon (now Bressanone) in 1450. Unfortunately, he could not take up his duties there for two years (the reason was opposition by the Duke of Austria), and the pope sent him as papal legate to North Germany and the Netherlands. His aim was to prepare the Christians for the Jubilee of 1450.

In Brixen, Nicholas began to reform the local Church, which caused him trouble. In 1460 he was imprisoned by the local ruler Sigismund. Set free he left his diocese and settled in Rome. He died in Todi in 1464. According to his wishes his body was buried in Rome and his heart in his home town Kues.

Nicholas' first important published work was *De docta ignorantia* (1440). This is perhaps his best known philosophical work. He argued there the incomplete nature of man's knowledge of the universe, claiming that the search for truth was equal to the task of squaring the circle. Among his writings on mathematics one should mention: *De geometricis transmutationibus* (1445), *De arithmeticis complementis* (1445), *De circuli quadratura* (1450), *Quadratura circuli* (1450), *De mathematicis complementis* (1453), *Dialogus de circuli quadratura* (1457), *De caesarea circuli quadratura* (1457), *De mathematica perfectione* (1458) and *Aurea propositio in mathematicis* (1459). He also wrote *Declaratio rectilineationis curvae* and *De una recti curvique mensura*, but their dates are unknown. He was interested in geometry and logic and had clearly made a study of at least parts of Euclid's *Elements* and the works of Thomas Bradwardine and Campanus of Novara. He contributed to the study of infinity, studying the infinitely large and the infinitely small. He looked at the circle as the limit of regular polygons and used it in his religious teaching to show how one can approach truth but never reach it completely. His main mathematical work is considered to be *De mathematicis complementi*. In many of his papers, he considered the problem of squaring the circle and of measuring the circumference of a circle.

He was also interested in astronomy. It led him to certain theories. Giordano Bruno is said to have written: "If [Nicholas of Cusa] had not been hindered by his priest's vestment, he would have even been greater than Pythagoras!".

In his philosophical works Nicholas was particularly interested in the theory of knowledge. He wrote on this topic in works such as *De conjecturis* (1440–1444) and *Compendium* (1464). According to him, knowledge is derived through the senses, but understanding is an abstraction of diverse sensory images. All human knowledge must be mere conjecture, and wisdom is attained only through understanding the extent of one's ignorance. After those general biographical remarks let us come to the proper subject, i.e. to Nicholas' philosophical views on mathematics. Note at the beginning that his writings on mathematics are those of a good amateur, and they do not attain top level in rigour.

Nicholas of Cusa, being convinced that human knowledge is only an approximation of the truth (*coniectura*), attributed to mathematical knowledge the highest degree of precision and clarity. Following the tradition of Boethius, he claimed that mathematics in the best way prepares the human mind for theological considerations. In *De docta ignorantia* [On Learned Ignorance] he wrote:

> Thus, Boethius, the most learned of the Romans, affirmed that anyone who altogether lacked skill in mathematics could not attain a knowledge of divine matters.[49] (I, 11)

And he added (*ibidem*):

> [...] since the pathway for approaching divine matters is opened to us only through symbols, we can make quite suitable use of mathematical signs because of their incorruptible certainty.[50]

Mathematics played an important, even fundamental, role in Nicholas' thought. In fact, it was for him an example and model of all veritable human knowledge. Mathematics gives the best possible certain and reliable knowledge. This is so because in mathematics the mind uses numbers and figures that are constructed by it without any reference to the knowledge of a changeable physical reality. In fact, numbers and figures are within the power only of the mind and emulate the activity of God – "And so, God, who created all things in number, weight, and measure[51], arranged the elements in an admirable order. (Number pertains to arithmetic, weight to music, measure to geometry.)[52]" (*De docta ignorantia* II, 13). And similarly at another place (*ibidem*):

> In creating the world, God used arithmetic, geometry, music, and likewise astronomy. (We ourselves also use these arts when we investigate the comparative relationships of objects,

---

49 „[...] ita ut Boethius, ille Romanorum litteratissimus, assereret neminem divinorum scientiam, qui penitus in mathematicis exercitio careret, attingere posse".

50 „ad divina non nisi per symbola accedendi nobis via pateat, quod tunc mathematicalibus signis propter ipsorum incorruptibilem certitudinem convenientius uti poterimus".

51 Wisd. 11, 21.

52 „Admirabili itaque ordine elementa constituta sunt per Deum, qui omnia in numero, pondere et mensura creavit. Numerus pertinet ad arithmeticam, pondus ad musicam, mensura ad geometriam".

of elements, and of motions.) For through arithmetic God united things. Through geometry He shaped them, in order that they would thereby attain firmness, stability, and mobility in accordance with their conditions.[53]

Any intellectual process presupposes the usage of numbers – in fact thinking means to count, to measure and to compare. Any human knowledge is expressed by numbers. Hence number is an indispensable stamp of human rationality.

According to the tradition of the Platonic Academy, Cusanus took up the classical tripartition of theoretical science: physics, mathematics and theology. Mathematical objects are intermediate between physical, material and changing realities and the reality that theology treats. Objects of mathematics – though more abstract than objects of sensual perception – are not free of any change. Still they are fixed and certain because they are in the power of the mind alone. He wrote (*De docta ignorantia* I, 11):

> In our considering of objects, we see that those which are more abstract than perceptible things, viz., mathematicals, (not that they are altogether free of material associations, without which they cannot be imagined, and not that they are at all subject to the possibility of changing) are very fixed and are very certain to us.[54]

The mind is internally bounded only by the principle of consistency.

At various places in his works, Nicholas mentioned the (not quite clear) idea of intellectual mathematics and physical mathematics – in particular he did so in connection with his considerations of the problem of squaring the circle. Physical mathematics is the inverted reflection of intellectual mathematics. The latter deals with the infinitely great and the infinitely small. It is the light of the mind – thanks to it one can do ordinary mathematics. Using Kant's terms one can say that intellectual mathematics is the condition that makes possible ordinary mathematics. Intellectual mathematics contains all the figures and forms that are distinct for reason. According to Nicholas, if the squaring of a circle is impossible on the level of ordinary mathematics, it exists on the level of the light of the intellect and of the superior mathematics. The latter can only be studied indirectly, on the basis of physical mathematics.

---

53 „Est autem Deus arithmetica, geometria atque musica simul et astronomia usus in mundi creatione, quibus artibus etiam et nos utimur, dum proportiones rerum et elementorum atque motuum investigamus. Per arithmeticam enim ipsa coadunavit; per geometriam figuravit, ut ex hoc consequerentur firmitatem et stabilitatem atque mobilitatem secundum condiciones suas [...]"

54 „Abstractiora autem istis, ubi de rebus consideratio habetur, – non ut appendiciis materialibus, sine quibus imaginari nequeunt, penitus careant neque penitus possibilitati fluctuanti subsint – firmissima videmus atque nobis certissima, ut sunt ipsa mathematicalia".

Where and how do mathematical objects exist? In his work *Idiota de mente* [The Layman on the Mind] he considered the concept of number. He distinguished numbers being the object of mathematics and numbers coming from God. The former come from man; the latter have their origin in God's mind. In *Idiota de mente* he wrote in Chapter 9:

> Mind makes a point to be the termination of a line, makes a line to be the termination of a surface, and makes a surface to be the termination of a material object. Mind makes number; hence, multitude and magnitude derive from mind. And, hence, mind measures all things.[55]

And in Chapter 6 one finds the following words:

> I deem the Pythagoreans – who, as you state, philosophize about all things by means of number – to be serious and keen [philosophers]. It is not the case that I think they meant to be speaking of number qua mathematical number and qua number proceeding from our mind. (For it is self-evident that that [sort of number] is not the beginning of anything.) Rather, they were speaking symbolically and plausibly about the number that proceeds from the Divine Mind – of which number a mathematical number is an image. For just as our mind is to the Infinite, Eternal Mind, so number [that proceeds] from our mind is to number [that proceeds from the Divine Mind]. And we give our name "number" to number from the Divine Mind, even as to the Divine Mind itself we give the name for our mind. And we take very great pleasure in occupying ourselves with numbers, as being an instance of our occupying ourselves with our own work.[56]

Hence the numbers being objects of mathematics are the image (*ymago*) of the numbers existing in God.

We see that Cusanus dissociates from Plato. For the latter, mathematical numbers belonged to a realm between the realm of ideas and the realm of objects sensually recognizable. They exist eternally. For Nicholas, there exist only numbers that come from God's mind and find their reality in the variety of sensually intelligible objects and the mathematical numbers that are creations of the human mind in accordance with God's numbers. He wrote in *Idiota de mente* (Chapter 6):

---

55 „Mens facit punctum terminum esse lineae et lineam terminum superficiei et superficiem corporis, facit numerum, unde multitudo et magnitudo a mente sunt, et hinc omnia mensurat".

56 „Arbitror autem viros Pythagoricos, qui ut ais per numerum de omnibus philosophantur, graves et acutos. Non quod credam eos voluisse de numero loqui, prout est mathematicus et ex nostra mente procedit – nam illum non esse alicuius rei principium de se constat –, sed symbolice ac rationabiliter locuti sunt de numero, qui ex divina mente procedit, cuius mathematicus est imago. Sicut enim mens nostra se habet ad infinitam aeternam mentem, ita numerus nostrae mentis ad numerum illum. Et damus illi numero nomen nostrum sicut menti illi nomen mentis nostrae, et delectabiliter multum versamur in numero *quasi* in nostro proprio opere".

You see, too, how it is that number is not anything other than the things enumerated. Here from you know that between the Divine Mind and things there is no actually existing intervening number. Instead, the number of things are the things.[57]

The role of numbers is seen by Cusanus in the following way (*Idiota de mente*, Chapter 6):

> In like manner, I say that number is the exemplar of our mind's conceptions. For without number mind can do nothing. If number did not exist, then there would be no assimilating, no conceptualizing, no discriminating, no measuring. For, without number, things could not be understood to be different from one another and to be discrete. For without number we [could] not understand that substance is one thing, quantity another thing, and so on regarding the other [categories]. Therefore, since number is a mode of understanding, nothing can be understood without it. For since our mind's number is an image of the divine number – which is the Exemplar-of-things – it is the exemplar of concepts.[58]

Add that according to Nicholas mathematical objects are good symbols of the essences of things. Hence, different kinds of reality can be symbolized by different kinds of numbers and unities. The divine unity should be symbolized by the first arithmetical unit – the one – which is the principle of all numbers. The unit of ten and its first multiples represent the order of pure intellects or intelligences, the unit of one hundred and its multiples – the order of souls, and the unit of one thousand can be linked to the world of bodies and materials.

In a similar way to numbers, Nicholas treats geometrical objects. They are creations of the human mind. In Chapter 9 of *Idiota de mente* he wrote:

> Mind makes a point to be the termination of a line, makes a line to be the termination of a surface, and makes a surface to be the termination of a material object [...].[59]

Also other geometrical figures like the circle, triangle, etc., are created by the mind (cf. *Idiota de mente*, Chapter 3):

---

57 „Conspicis etiam, quomodo non est aliud numerus quam res numeratae. Ex quo habes inter mentem divinam et res non mediare numerum, qui habeat actuale esse, sed numerus rerum res sunt".

58 „Pariformiter dico exemplar conceptionum nostrae mentis numerum esse. Sine numero enim nihil facere potest; neque assimilatio neque notio neque discretio neque mensuratio fieret numero non exsistente. Res enim non possunt aliae et aliae et discretae sine numero intelligi. Nam quod alia res est substantia et alia quantitas et ita de aliis, sine numero non intelligitur. Unde cum numerus sit modus intelligendi, nihil sine eo intelligi potest. Numerus enim nostrae mentis cum sit imago numeri divini, qui est rerum exemplar, est exemplar notionum".

59 „Mens facit punctum terminum esse lineae et lineam terminum superficiei et superficiem corporis [...]".

You know, O Orator, how it is that we produce mathematical figures by the power of our mind. Hence, when I wish to make triangularity visible, I construct a figure in which I make three angles, so that, thereupon, triangularity shines forth in the figure thus arranged and proportioned. To triangularity is united a name, which, by imposition, is "trigon". Accordingly, I say: if "trigon" were the precise name of the triangular figure, then I would know the precise names of all polygons. For, in that case, I would know that the name of a quadrangular figure ought to be "tetragon" and that the name of a fiveangled figure ought to be "pentagon", and so on. And from a knowledge of the one name I would know (1) the figure named, (2) all nameable polygons, (3) their differences and agreements, and (4) whatever else could be known in regard to this matter.[60]

How does a human mind create a geometrical object? Cusanus explains this in Chapter 9 of *Idiota de mente* writing:

> *Philosopher*: How does the mind make a line?
> *Layman*: By considering length without width. And [mind makes] a surface by going on to consider width without solidity. (However, neither a point nor a line nor a surface can actually exist in this way, for outside the mind only solidity actually exists.) Thus, the measure or end-point of each thing is due to mind. Stones and pieces of wood have a certain measurement – and have endpoints – outside our mind; but these [measurements and end-points] are due to the Uncreated Mind, from which all the end-points of things derive.[61]

In *De docta ignorantia* (II, 5) he adds:

> In order that you may see more clearly: a line cannot exist actually except in a material object, as will be shown elsewhere.[62]

What does it mean to exist actually? Nicholas explains it in the following way (*De docta ignorantia* II, 5):

---

60 „Tu nosti, orator, quomodo nos exserimus ex vi mentis mathematicales figuras. Unde dum triangularitatem visibilem facere voluero, figuram facio, in qua tres angulos constituo, ut tunc in figura sic habituata et proportionata triangularitas reluceat, cum qua unitum est vocabulum, quod ponatur esse «trigonus». Dico igitur: Si «trigonus» est praecisum vocabulum figurae triangularis, tunc scio praecisa vocabulo omnium polygoniarum. Scio enim tunc, quod figurae quadrangularis vocabulum esse debet «tetragonus» et quinquangularis «pentagonus» et ita deinceps. Et ex notitia nominis unius cognosco figuram nominatam et omnes nominabiles polygonias et differentias et concordantias earundem et quidquid circa hoc sciri potest".

61 „PHILOSOPHUS: Quomodo facit lineam? IDIOTA: Considerando longitudinem sine latitudine, et superficiem considerando latitudinem sine soliditate, licet sic actu nec punctus nec linea nec superficies esse possit, cum sola soliditas extra mentem actu exsistat. Sic omnis rei mensura vel terminus ex mente est. Et ligna et lapides certam mensuram et terminos habent praeter mentem nostram, sed ex mente increata, a qua rerum omnis terminus descendit".

62 „Et ut clarius videas: Linea actu esse nequit nisi in corpore, ut ostendetur alibi".

But everything which exists actually, exists in God, since He is the actuality of all things. Now, actuality is the perfection and the end of possibility.[63]

Mathematical objects created by the human mind are a picture (*ymago*) of that which comes from God's mind and is realized in things. Those mathematical objects can be made by the mind thanks to its ability of assimilation – see the subtitle of Chapter 7 of *Idiota de mente* that says:

> CHAPTER SEVEN: Mind produces from itself, by means of assimilation, the forms of things; and it attains unto absolute possibility, or matter.[64]

At other places Nicholas uses instead of "assimilation" the word "abstraction" (cf. *De docta ignorantia* II, 1 and 4). One can see here a form of empiricism. In fact in *Idiota de mente* (Chapter 2) he writes[65]:

> So whoever thinks that in the intellect there can be nothing that is not present in reason also thinks that in the intellect there can be nothing that was not first in the senses.[66]

Let us turn now to Cusanus' views concerning infinity. It appears by him both in mathematical considerations as well as in his philosophico-theological considerations. He claims that infinity can be grasped in mathematics by mind with the help of concepts, but it cannot be grasped with the help of senses. It should be stressed that the reason and the aim for considering infinity in mathematics was for Nicholas an attempt to approach the infinity of God.

Considering the problem of the applicability of the Aristotelian category of quantity, Nicholas argued that infinity cannot be characterized in terms of this category; it cannot be quantified. Such notions as "bigger" or "smaller", "equal" or "unequal" cannot be related to infinity. Human rationality operates epistemologically within the category of quantity; all mathematical operations are based on it. Hence there are some constraints put on mathematics in its reach for infinity. In fact, there is no way from quantity to infinity. Such notions as infinity, maximum or minimum are all transcendent terms. Cusanus objected to Aristotle's idea of potential infinity because it is based on an infinite progression of finite quantities. Infinity cannot be measured. On the other hand, it is the measure of everything else; and it is unique. Infinity defies also any logical treatment. The infinite has no proportion to the finite, hence it will never be known from the finite.

---

63 „Omne autem actu existens in Deo est, quia ipse est actus omnium. Actus autem est perfectio et finis potentiae".
64 „Quomodo mens a se exserit rerum formas via assimilationis et possibilitatem absolutam seu materiam attingit".
65 Add that intellect was by Nicolas the higher mental faculty.
66 „Quicumque igitur putat nihil in intellectu cadere posse, quod non cadat in ratione, ille etiam putat nihil posse esse in intellectu, quod prius non fuit in sensu".

Mathematics can help us to understand infinity, in particular God's infinity. This can be done by symbolic illustration. In *De docta ignorantia* (I, 12) he wrote:

> For since all mathematicals are finite and otherwise could not even be imagined: if we want to use finite things as a way for ascending to the unqualifiedly Maximum, we must first consider finite mathematical figures together with their characteristics and relations. Next, [we must] apply these relations, in a transformed way, to corresponding infinite mathematical figures. Thirdly, [we must] thereafter in a still more highly transformed way, apply the relations of these infinite figures to the simple Infinite, which is altogether independent even of all figure. At this point our ignorance will be taught incomprehensibly how we are to think more correctly and truly about the Most High as we grope by means of a symbolism.[67]

Among things and processes that can be known by the senses there is nothing that could not be increased and expanded. Hence infinity cannot be realized in any process. On the other hand, in mathematics there are examples showing that the limit of a process can be grasped by a concept. As such an example Nicholas gives a sequence of regular polygons of n sides. If n grows unboundedly, then the polygons approximate better and better a circle. Among objects cognizable by the senses there exists no circle. a circle exists only as a concept in our mind. In *Idiota de mente* (Chapter 7) he wrote:

> [...] as, for example, when it conceives a circle to be a figure from whose center all lines that are extended to the circumference are equal. In this way of existing no circle can exist extra-mentally, in matter.[68]

Such different objects as a circle and a regular polygon of n sides coincide in infinity. More similar examples can be found by Nicholas. In Chapter 13 of *De docta ignorantia* he writes about a sequence of circles that are tangent to a given line at one fixed point and whose radius grows to infinity. The limit of such a sequence can be grasped as a concept – namely by the concept of a line. According to Cusanus, different geometrical figures (circles, spheres, lines, triangles) can be identified with one another when they are increased to the infinite. In particular, the infinite circle and the infinite line can be identified.

---

67 „Nam cum omnia mathematicalia sint finita et aliter etiam imaginari nequeant: si finitis uti pro exemplo voluerimus ad maximum simpliciter ascendendi, primo necesse est figuras mathematicas finitas considerare cum suis passionibus et rationibus, et ipsas rationes correspondenter ad infinitas tales figuras transferre, post haec tertio adhuc altius ipsas rationes infinitarum figurarum transumere ad infinitum simplex absolutissimum etiam ab omni figura. Et tunc nostra ignorantia incomprehensibiliter docebitur, quomodo de altissimo rectius et verius sit nobis in aenigmate laborantibus sentiendum".

68 „[...] dum concipit circulum esse figuram, a cuius centro omnes lineae ad circumferentiam ductae sunt aequales, quo modo essendi circulus extra mentem in materia esse nequit".

In a similar way the concept of a line cannot be realized in a world of objects known by the senses. He comes to the conclusion (*De docta ignorantia* I, 13):

> I maintain, therefore, that if there were an infinite line, it would be a straight line, a triangle, a circle, and a sphere. And likewise if there were an infinite sphere, it would be a circle, a triangle, and a line. And the same thing must be said about an infinite triangle and an infinite circle.[69]

In all these cases Cusanus talks about *coincidencia oppositorum*. He treats it as a principle and applies it not only in mathematics but also in non-mathematical domains where an unlimited object is never given but can be grasped only by finite approximations.

The completion of a process (and simultaneously its limit) have for Cusanus the highest form of being and is eternal because the process itself seeks its own completion.

Considering a line he writes in connection with this in *De venatione sapientia* [The Hunt for Wisdom] (Chapter 34):

> To this end, I draw a line a b, and I say that the line a b is great, because it is greater than one half of itself, and that it can be made greater by extending, or augmenting, it. But it will not become a greatness which, since [it cannot be made greater], would be what it can be. If a line were made so great that it could not be greater, it would be that which it could be; and, [in that case], it would not be made but would be eternal and would precede the possibility-of being-made and would not be a line but would be Eternal Greatness. In the foregoing way I see that since whatever can be made greater is subsequent to the possibility-of-being-made, it is never made to be [all] that which it can be. But because Greatness is [all] that which it can be, it cannot be either greater or lesser [than it is]. And so, Greatness is neither greater nor lesser than anything great or than anything small but is the efficient Cause of all things great or small, and is their formal Cause and final Cause and their most adequate Measure. In all great things and all small things Greatness is all [these] things; and, at the same time, it is none of all [these] things, since all great things and all small things are subsequent to the possibility-of-being-made, which Greatness precedes.

The infinite does not borrow its existence from finite objects. The finite cannot guarantee the existence of the infinite because the latter will never be reached in a process of approximation by finite elements. Just the opposite – the infinite is first, and remains in the order of existence ahead of all that is finite. Cusanus reverses here the order of thinking. According to him the finite can be understood and grasped only with the help of the infinite. In *Idiota de mente* he wrote (Chapter 2):

---

69 „Dico igitur, quod, si esset linea infinita, illa esset recta, illa esset triangulus, illa esset circulus et esset sphaera; et pariformiter, si esset sphaera infinita, illa esset circulus, triangulus et linea; et ita de triangulo infinito atque circulo infinito idem dicendum est".

Consequently, everything finite is originated from the Infinite Beginning.[70]

A finite segment is imperfect in comparison with an infinite line. In *De venatione sapientiae* (Chapter 26) he wrote:

> But since there is no line that is without a length, a line that is not as long as its length [could be] is imperfect in comparison with a line that cannot be longer.

In a similar way he wrote in *De docta ignorantia* (II, 5):

> Now, every finite line has its being from the infinite line, which is all that which the finite line is. Therefore, in the finite line all that which the infinite line is – viz., line, triangle, and the others – is that which the finite line is.[71]

One can see that the idea of *coincidentia oppositiorum* that Cusanus used in his attempts to explain how our (mathematical) knowledge can approach God's knowledge is now applied by him as a principle of the ontology of mathematics. And he is doing so quite consciously. Indeed, in *De mathematica perfectione* he writes: "My aim is to improve mathematics by *concidentia oppositorum*"[72]. In this work, Cusanus used this concept as a tool for creating the new mathematical procedure of infinite approximation. He tried namely to calculate the circumference of a circle – in this way infinity had become by him a methodological tool. It was also the reflection of his understanding of epistemology as an approximate process towards the truth. One can see in it the creation of the epistemological prerequisites of modern natural science.

---

70 „Quare omne finitum principiatum ab infinito principio".
71 „Omnis autem linea finita habet esse suum ab infinita, quae est omne id, quod est. Quare in linea finita omne id, quod est linea infinita (ut est linea, triangulus, et cetera), est id, quod est linea finita".
72 „Intentio est ex oppositorum coincidentia mathematicam venari perfectionem".

Phenomenological Ideas in the Philosophy of Mathematics. From Husserl to Gödel

Co-authored by Thomas Bedürftig

Husserl's philosophy of mathematics

Husserl came in a certain sense from mathematics. He began his studies of mathematics at the universities of Leipzig and Berlin with Carl Weierstraß and Leopold Kronecker. In 1881 he moved to Vienna where he studied with Leo Königsberger and in 1883 obtained his doctor's degree on the base of the dissertation *Beiträge zur Variationsrechnung*. Strongly impressed by the lectures of Franz Brentano (1838–1917) on psychology and philosophy which he attended at the University of Vienna, he decided after the doctorate to dedicate his life to philosophy. In 1886 he went to the University of Halle to obtain his Habilitation with Carl Stumpf, a former student of Brentano. The *Habilitationsschrift* was entitled *Über den Begriff der Zahl. Psychologische Analysen*. This 64-page work was later expanded into a book (of five times the length), which was one of Husserl's major works: *Philosophie der Arithmetik. Psychologische und logische Untersuchungen* (Husserl 1891, cf. also Husserl 1970, 2003).

Working as *Privatdozent* at the University of Halle, Husserl came into contact with mathematicians: Georg Cantor, the founder of set theory and Hermann Grassmann's son, also Hermann. The former, with whom he had long philosophical conversations when they were teaching together in Halle in the 1890s, told him about Bernard Bolzano. In fact, Husserl was perhaps the first philosopher outside Bohemia to be influenced significantly by Bolzano (cf. Grattan-Guinness 2000). Later, as a professor of the University in Göttingen, Husserl had contact with David Hilbert and as a professor in Freiburg (Breisgau), where he was appointed in 1916, with Ernst Zermelo.

Cantor influenced in a certain sense the earlier works of Husserl though he is quoted only twice in Husserl's Habilitationsschrift. Similarly, discussions with Gottlob Frege – the founder of logicism, one of the main trends in the modern philosophy of mathematics – influenced him.[73] Both Cantor and Frege will appear below when we shall describe Husserl's philosophy of arithmetic. Considering the connections of Husserl with mathematics and mathematicians, one can say that

---

73 For trends in the philosophy of mathematics see, e.g., Bedürftig and Murawski (2015, 2018).

his philosophy had, in fact, no visible meaning for the mathematics of his time; however, on the contrary, mathematics strongly influenced his philosophy.

One of the mathematical motives of Husserl's philosophy can be recognized in Weierstraß's program of arithmetization of analysis. Its aim was to found the whole of mathematics on the base of arithmetic, and to define all its concepts in terms of arithmetical ones. Quite a lot of mathematicians of the 19th century initialized and supported this arithmetization, among them Augustin-Louis Cauchy, Bernard Bolzano, Richard Dedekind, Georg Cantor and Carl Weierstraß himself. Husserl's aim was to justify the Weierstraß program by his investigations, philosophically and psychologically. In the Preface to *Philosophie der Arithmetik* he wrote (Husserl 1891, p. VIII): "Perhaps my efforts should not be wholly worthless, perhaps I have succeeded in preparing the way, at least on some basic points, for the true philosophy of the calculus, that desideratum of centuries".

Aspects of phenomenological methods and basic concepts of phenomenology can already be seen in Husserl's *Habilitationsschrift*. The latter, as well as his book *Philosophie der Arithmetik*, was influenced by Brentano and stamped by his descriptive psychology. Later Husserl moved away from this "psychologism" and criticized the psychological point of view in the philosophy of logic and mathematics – for example in the first volume of his *Logische Untersuchungen* (1900–1901).[74] He was of the opinion that phenomenological data are correctly described by empirical psychology. He changed his mind around 1930 claiming now that there is in fact no direct connection and that psychological analysis cannot be used in phenomenology. This purely philosophically and *a priori* treated phenomenology that should remove psychology as its foundation was developed by Husserl for more than 40 years. His aim was to establish philosophy as a strict science and to create the universal foundation of all disciplines.

Mathematics was for Husserl a typical example of an eidetic discipline. According to him, mathematics studies the fundamental objects, like numbers in the arithmetic and forms or similar phenomena in the geometry.[75] Husserl claimed that one can penetrate in a kind of *Wesensschau* to their essences, their *eidos* – as

---

74 However, some forms of psychologism which he analysed there and tried to reject can be seen not directly in his *Philosophie der Arithmetik*. There are, however, some concepts that appear and are considered both in *Philosophie der Arithmetik* and in *Logische Untersuchungen*, but they are treated in a different way. For example, in *Philosophie der Arithmetik* an important role is played by the concept of abstraction taken from the psychological point of view. The same term is present in *Logische Untersuchungen* together with a sophisticated theory and many possible variants.

75 It is worth noting here that Husserl proposed an extension of geometry in the direction called today "topology".

in the case of physical objects. He made here no difference. Mathematics, as an eidetic discipline, studies abstract objects in which intentionally is more than we can recognize in our normal cognition and to which we will be phenomenologically led back.

Husserl was not satisfied with the solutions of the program of arithmetization proposed by Dedekind, Cantor and others. His own position, especially in *Philosophie der Arithmetik*, was resolutely antiaxiomatic. According to him, one should not found "arithmetic on a sequence of formal definitions, out of which all the theorems of that science could be deduced purely syllogistically" – as he wrote in *Philosophie der Arithmetik* (1891, p. 130; 2003, p. 127). As soon as one comes to the ultimate, elementary concepts, the whole process of defining has to come to an end and one should point to the concrete phenomena from or through which the concepts are abstracted and to show the nature of the abstraction process.

He wrote (Husserl 2003, pp. 310–311):

> Today there is a general belief that a rigorous and thoroughgoing development of higher analysis [...] excluding all auxiliary concepts borrowed from geometry, would have to emanate from elementary arithmetic alone, in which analysis is grounded. But this elementary arithmetic has, as a matter of fact, its sole foundation in the concept of number; or, more precisely put, it has it in that never-ending series of concepts which mathematicians call "positive whole numbers". [...] Therefore, it is with the analysis of the concept of number that any philosophy of mathematics must begin.

In *Philosophie der Arithmetik*, Husserl referred, as mentioned above, to Brentano's method of descriptive psychology and understood – similarly to Weierstraß and other mathematicians of that time – natural numbers by empirical counting, what by him is masked by other principles. In the first part of the work, Husserl developed a psychological analysis that started from the everyday concept of a number. The analysis begins with the development, application and appearance of numbers; and on this base he tries to explain the psychological origin of numbers. He claims that the fundamental concept of a number cannot be defined (Husserl 1891, p. 142; 2003, p. 136):

> [...] the difficulty lies in the phenomena, in their correct description, analysis and interpretation. It is only with reference to the phenomena that insight into the essence of the number concept is to be won.

These words exhibit Husserl's psychological belief from this period. We find here already the "reference to the phenomena".

Since our intellect and time are bounded, we are able to achieve the comprehension only of a very small part of mathematics. In order to overcome those limits one introduces symbols which accompany and guide our thinking. Almost all we know about arithmetic we know indirectly via the intermediation of symbols. This explains why in the second part of *Philosophie der Arithmetik* Husserl considers extensively the symbolic representations.

As indicated above Husserl – being against the axiomatic approach to the characterization of numbers – claimed that the challenge is to find the sources of the number concept, to comprehend the nature of the abstraction process and to describe the concept formation. According to that one should focus on "our grasp of the concept of number" and not on the number as such.

Husserl understands abstraction in the following way: "to abstain from something or abstract from something means simply: not notice this especially". And he explains: abstraction "does not have the effect that its content and its connections disappear from our consciousness" (Husserl 1891, p. 85; 2003, p. 83). It is here psychologically indicated what Husserl later included into his method of phenomenological reduction. That there are contents that are "not especially noticed" – just they make possible the *Colligieren*, the connecting to a new whole.

This *Colligieren*, which leads to "multiplicities", is for Husserl directly connected with the concept of number. This is one of two principles that are fundamental for numbers. The second principle is the principle of "something" underlying everything. "The 'something' is no abstract partial content" of any "concrete multiplicity". For "the concept of something is due to the reflection on the psychic act of conception" (Husserl, 1891, p. 86; 2003, p. 84). Again one can suppose that here – psychological, intentional something – presages the later philosophical *eidos*.

By such copies of something general, multiplicities are constituted: "A multiplicity is nothing more than: something and something and something etc.; or any one and any one and any one etc.; or briefly: one and one and one etc." (Husserl 1891, p. 85; 2003, p. 83). In the word "one" Husserl sees the relation of "partial content" with the whole of the multiplicity that is not expressed in "something".

Multiplicity and quantity (*Anzahl*) – and here we are at Husserl's concept of number – can be hardly distinguished. "It is *a priori* apparent that they coincide in their essential content" (Husserl 1891, p. 89). "Quantity" is the "generic term": the concept of quantity distinguishes the "abstract forms of multiplicities", cancels the "vague indefiniteness" of multiplicities and appends to them the "sharply definite how many" (*loc. cit.*). Multiplicity for Husserl resembles the "something" of number, an indefinable psychological datum (cf. Husserl 1891, p. 130; 2003, p. 127).

The essential element of the abstraction that leads to the just mentioned concept of quantity is in the concept of "something" (cf. Husserl 1891, p. 129; 2003, p. 128). It spares – differently as in the case of the set-theoretical concept of a cardinal number – the comparison of "concrete multiplicities", which Husserl explicitly notices. Husserl's concept of quantity comes back to the age-old "definition" of a number by Euclid and corresponds to Cantor's characterization of cardinal numbers stating that it is "a definite aggregate composed of units" (Cantor 1895,

p. 482). We recognize that with Husserl one has to do here with a process which goes like a counting: "One and one and one etc.". He treats numbers as arising and given additively, for example, "three" as "one, one and one" (p. 87). The counting process is so explicit and clear in this formulation that it seems that Husserl does not separate quantity and counting. At least in such a way he articulates it in his (very sharp and not consistent) critique of Kant's concept of schemata[76] (Husserl 1891, p. 86; 2003, p. 84):

> Number is the idea of a universal procedure of imagination getting the concept of quantity an image. However this procedure can only mean counting. But is it not clear that "number" and the idea of "counting" are the same?

This remark is a bit surprising because just at that time Dedekind (1888) and Peano (1889) provided a clear mathematical separation of number and quantity, the grounding of the concept of number by counting. It seems as if the mathematician Husserl did not want to notice this in his psychological, anti-axiomatic attitude. One can briefly characterize Husserl's concept of number by saying that, according to him, numbers are quantities; and quantities are distinguished multiplicities of abstract units.

Just these quantities are for Husserl primary for the concept of number. On the other hand, cardinal numbers as classes of equipollent sets are unfinished and "useless concept formations" (Husserl 1891, p. 129) which state no number, but only the equality of number or quantity. This "definition" (Husserl himself puts this in quotation marks) is "considerably appreciable" (Husserl 1891, p. 130 f.) only for "this Wildman" on "that level of mind" for whom the symbolic counting is not available.

In our opinion, Husserl misunderstands both Cantor (in favour of him) as well as Frege, and finally also Dedekind. Note that he criticizes only the decided antipsychologist Frege.

According to Husserl, a mathematician operates not with abstract numbers but with quantities that are always connected with the idea of special sets via multiplicities.

Mathematics itself is for Husserl a formal ontology. Objects investigated by mathematics are formal categories in various forms – and they are themselves not perceivable. Numbers are here an example. Thanks to the ability of categorical abstraction we can free ourselves from the empirical components of judgements and concentrate ourselves on the formal categories. In the eidetic intuition and variation we are able to grasp the possibility, impossibility, necessity and contingency of connections between concepts or between formal categories. The categorical abstraction and the eidetic intuition form the base of the mathematical knowledge.

---

76 Cf. Kauferstein (2006, p. 108 ff.)

Comparing Husserl and Frege one sees that for the former a direct experience, i.e. perception, is the ultimate basis for the meaningful analysis of numbers (and other mathematical notions), whereas the latter relies on the certainty given by logic. Husserl wants only to describe our experiences. Frege's logical analysis consists in constructing a notion of number in the ideography. For Husserl, such an approach is artificial or, as he says, "chimerical" (cf. Husserl 1970, pp. 119–120; 2003, p. 125). He claims that one should analyse concepts as they are given to us.

Weyl's and Becker's phenomenological philosophy of mathematics

The ideas of Husserl found response in papers of the famous German mathematician Hermann Weyl (1885–1955). His interests in philosophy go back to his graduate student days between 1904 and 1908, and his allegiance to it lasted till the early 1920s. a few years after the publication of *Das Kontinuum* (1918), Weyl joined the intuitionistic camp of L.E.J. Brouwer and developed his own approach to intuitionism, claiming that philosophy and intuitionism are strongly connected. Later, Weyl changed his views again and legitimated Hilbert's program. All this was connected with his critique of phenomenology. Mancosu and Ryckman (2005, p. 242) claim that "[a]pparently failing to discriminate between the resources available to phenomenology and those of intuitionistic mathematics in accounting for a contentual Anschauung capable of grounding the meaning of mathematical statements, Weyl saw the failure of the latter, in the face of Hilbert's finitism, as implicating the failure of the former". Weyl wrote (1967, p. 484; original emphasis):

> If Hilbert's view prevails over intuitionism, as appears to be the case, *then I see in this a decisive defeat of the philosophical attitude of pure phenomenology*, which thus proves to be insufficient for the understanding of creative science even in the area of cognition that is most primal and most readily open to evidence – mathematics.

The influence of Husserl's ideas on Weyl can be seen in the care with which he treated issues like the relationship between intuition and formalization (cf. Weyl 1918), the connection between his construction postulates and the idea of a pure syntax of relations, the appeal to a *Wesensschau*, etc. In the Preface to the work *Das Kontinuum*, Weyl explicitly declares that he agrees with the conceptions that underlie Husserl's *Logische Untersuchungen* with respect to the epistemological side of logic. Answering Husserl's gift of the second edition of *Logische Untersuchungen* to him and his wife, he wrote in a letter to Husserl (cf. Husserl 1994, p. 290):

> You have made me and my wife very happy with the last volume of the *Logical Investigations*; and we thank you with admiration for this present. [...] Despite all the faults you attribute to the *Logical Investigations* from your present standpoint, I find the conclusive results of this work – which has rendered such an enormous service to the spirit of pure

objectivity in epistemology – the decisive insights on evidence and truth, and the recognition that "intuition" [*Anschauung*] extends beyond sensual intuition, established with great clarity and conciseness.

On the other hand, Husserl read Weyl's *Das Kontinuum* as well as his *Raum, Zeit, Materie* (1922), and found them close to his views. He stressed and praised Weyl's attempts to develop a philosophy of mathematics on the base of logico-mathematical intuition. Husserl was pleased to have Weyl – who was a prominent mathematician – on his side. In a private correspondence he wrote to Weyl that his works were being read very carefully in Freiburg and had had an important impact on new phenomenological investigations, in particular those of his assistant – Oskar Becker.

Oskar Becker (1889–1964) studied mathematics at Leipzig and wrote his doctoral dissertation in mathematics under Otto Hölder and Karl Rohn in 1914. He then devoted himself to philosophy and wrote his *Habilitationsschrift* on the phenomenological foundations of geometry and relativity under Husserl's supervision in 1923. He admitted that it was Weyl's work that made a phenomenological foundation of geometry possible. Becker became Husserl's assistant in the same year. In 1927, he published his major work *Mathematische Existenz* (1927). The book was strongly influenced by Heidegger's investigations, in particular by his investigations on the facticity of *Dasein*. This led Becker to pose the problem of mathematical existence within the confines of human existence. He wrote: "The factual life of the mankind [...] is the ontical foundation also for the mathematical" (Becker 1927, p. 636). This standpoint in the philosophy of mathematics led Becker to find the origin of mathematical abstractions in concrete aspects of human life. In this way he became critical of Husserl's style of phenomenological analysis. This anthropological current played an important role in Becker's analysis of the transfinite. Hence, Becker utilized not only Husserl's phenomenology but also Heideggerian hermeneutics, in particular discussing the infinity of arithmetical counting as "being towards death" (*Sein-zum-Tode*).[77]

At the end of his life, Becker re-emphasized the distinction between intuition of the formal and Platonic realm as opposed to the concrete existential realm and developed his own approach to the phenomenology called by him *mantic*. With this word he referred to the fact that there is a divinatory aspect related to any attempt

---

77 Note that being (since 1923) an assistant of Husserl, Becker was attending seminars by Heidegger. This can explain the influence of the latter on Becker's *Mathematische Existenz*. Add that *Mathematische Existenz* and Heidegger's *Sein und Zeit* were published in 1927 in the same issue of *Jahrbuch für Philosophie und Phänomenologische Forschung*.

to understand *Natur*. In the light of this mathematics appears as a divinatory science which by means of symbols allows us to go beyond what is accessible. Mantic phenomenology will have to replace the older "eidetic" phenomenology.

Becker's works have not had great influence on later debates in the foundations of mathematics, despite the many interesting analyses included in them, in particular of the existence of mathematical objects.

Talking about Weyl and Becker one should mention also Felix Kaufmann (1895–1949), an Austrian-American philosopher of law. He studied jurisprudence and philosophy in Vienna, and from 1922 till 1938 (when he left for the USA) he was a *Privatdozent* there. He was associated with the Vienna Circle. He wrote on the foundations of mathematics attempting, along with Weyl and Becker, to apply the phenomenology of Husserl to constructive mathematics. His main work here is the book *Das Unendliche in der Mathematik und seine Ausschaltung* (1930).

Gödel's philosophy of mathematics *versus* phenomenology

One of the most eminent logicians and philosophers of mathematics in whom we find Husserl's phenomenological ideas is Kurt Gödel (1906–1978). Let us start by noting that Husserl never referred to Gödel. In fact he was more than 70 years old when Gödel obtained his great results on incompleteness and consistency, and he died a few years later, in 1938. It is claimed, however (cf. Hartimo 2017), that he knew of Gödel's results. Also Gödel never referred to Husserl in his published works. However, his *Nachlass* shows that he knew Husserl's work quite well and appreciated it highly.

Gödel started to study Husserl's works in 1959 and became soon absorbed by them finding the author quite congenial. He owned all Husserl's main works.[78] The underlinings and comments (mostly in Gabelsberger shorthand) in the margin indicate that he studied them carefully. Most of his comments are positive and expand upon Husserl's points, but sometimes he is critical. One should note that Gödel expressed philosophical views on mathematics similar to those of Husserl long before he started to study them (cf. Føllesdal 1995, p. 428). Views found in Husserl's writings were not radically different from his own. It seems that what impressed him was Husserl's general philosophy which would provide a systematic framework for a number of his own earlier ideas on the foundations of mathematics. Hao Wang (1996, p. 166) writes that "Gödel's own main aim in philosophy was to develop metaphysics – especially, something like the monadology of Leibniz transformed into exact theory – with the help of phenomenology".

---

78 He owned among others *Logische Untersuchungen* (in the edition from 1968), *Ideen*, *Cartesianische Meditationen und Pariser Vorträge*, *Die Krisis der europäischen Wissenschaften und die transzendentale Phänemenologie*.

Gödel considered both central questions in the philosophy of mathematics: (1) What is the ontological status of mathematical entities, and (2) How do we find out anything about them? Considering the first problem, one should say that Gödel had held realist views on mathematical entities since his student days (cf. Wang 1974, pp. 8–11) – more exactly since 1921-1922. In *Russell's mathematical logic* he wrote about classes and concepts (Gödel 1944):

> It seems to me that the assumption of such objects is quite as legitimate as the assumption of physical bodies and there is quite as much reason to believe in their existence. They are in the same sense necessary to obtain a satisfactory systems of mathematics as physical bodies are necessary for a satisfactory theory of our sense perceptions and in both cases it is impossible to interpret the propositions one wants to assert about these entities as propositions about the "data", i.e., in the latter case the actually occurring sense perceptions.

Similar views were expressed by him in his Gibbs lecture (1951) and in the unfinished contribution to the book *The Philosophy of Rudolf Carnap* titled *Is mathematics syntax of language?* (Gödel 1953). He writes there about concepts and their properties (Gödel 1953, p. 9):

> Mathematical propositions, it is true, do not express physical properties of the structures concerned [in physics], but rather properties of the *concepts* in which we describe those structures. But this only shows that the properties of those concepts are something quite as objective and independent of our choice as physical properties of matter. This is not surprising, since concepts are composed of primitive ones, which, as well as their properties, we can create as little as the primitive constituents of matter and their properties.

It should be stressed that Gödel does not claim here the objective existence of properties, but says only that they are as objective as the physical properties of matter. Compare this with Husserl's claim that abstract objects of mathematics have – like other essences – the same ontological status as physical objects, that they are objective, but not in the straightforward realist sense (cf. Føllesdal 1995, p. 432 and 439).

The comparison of the status of mathematical objects and physical objects one finds also in Supplement to the second edition of Gödel's paper *What is Cantor's Continuum Problem?* where he says (cf. Gödel 1947/1964, p. 272) that the question of the objective existence of the objects of mathematical intuition is an exact replica of the question of the objective existence of objects of the outer world. Føllesdal (1995, p. 440) notes that "Gödel's use of the phrase 'exact replica' brings to mind the analogy Husserl saw between our intuition of essences in *Wesensschau* and of physical objects in perception".

Let us turn now to the second problem, i.e. to the epistemology of mathematics. As indicated above in *Russell's mathematical logic* (1944), Gödel talked about elementary mathematical evidence or mathematical "data" and compared it to sense perception. The notion of mathematical intuition was also discussed by him in the papers (1951) and (1953) quoted above. (1951, p. 320) he writes:

> What is wrong, however, is that the meaning of the terms (that is, concepts they denote) is asserted to be something manmade and consisting merely in semantical conventions. The truth, I believe, is that these concepts form an objective reality of their own, which we cannot create or change, but only perceive and describe.

In (1953, p. 359) he writes:

> The similarity between mathematical intuition and physical sense is very striking. It is arbitrary to consider "This is red" an immediate datum, but not so to consider the proposition expressing modus ponens or complete induction (or perhaps some simpler propositions from which the latter follows). For the difference, as far as it is relevant here, consists solely in the fact that in the first case a relationship between a concept and a particular object is perceived, while in the second case it is a relationship between concepts.

In the Supplement to the second edition of *What is Cantor's Continuum Problem?* he writes (Gödel 1947/1964, p. 271):

> But despite their remoteness from sense experience, we do have something like a perception of the objects of set theory, as is seen from the fact that the axioms force themselves upon us as being true. I don't see any reason why we should have less confidence in this kind of perception, i.e., in mathematical intuition, than in sense perception, which induces us to build up physical theories and to expect that future sense perceptions will agree with them [...].

Gödel did not explain what is the object of mathematical intuition. There are the following possibilities: propositions (cf. Gödel 1953), concepts (cf. Gödel 1951), sets and concepts (Gödel 1947/1964), or all three. Recall that Husserl distinguished two kinds of intuition: perception (where physical objects are intuited) and eidetic intuition (where the object is an eidetic entity or a "something" according to his *Philosophie der Arithmetik*) and claimed that the latter is more basic. It is not clear whether Gödel shared his views in this respect.

It is worth quoting still one passage from the second edition (Gödel 1947/1964) where he wrote (p. 271):

> That something besides the sensations actually is immediately given follows (independently of mathematics) from the fact that even our ideas referring to physical objects contain constituents qualitatively different from sensations or mere combinations of sensations, e.g., the idea of object itself. [...] Evidently the "given" underlying mathematics is closely related to the abstract elements contained in our empirical ideas. It by no means follows, however, that the data of this second kind, because they cannot be associated with actions of certain things upon our sense organs, are something purely subjective, as Kant asserted. Rather they, too, may represent an aspect of objective reality, but, as opposed to the sensations, their presence in us may be due to another kind of relationship between ourselves and reality.

Føllesdal (1995, p. 442) suggests that Gödel's point in this passage is that "what is given in our experience is not just physical objects, but also various abstract features that are instantiated by these objects".

Mathematical intuition cannot guarantee us certainty of our knowledge. In fact neither perception nor categorical intuition are infallible sources of evidence. Gödel writes about four different methods one can use to get insight into mathematical reality:

- elementary consequences,
- success, i.e. fruitfulness in consequences,
- clarification and
- systematicity.

The first one is involved in the situation when recondite axioms have elementary consequences, e.g. axioms concerning great transfinite numbers can have consequences in the arithmetic of natural numbers. Clarification refers to situations when a discussed hypothesis cannot be solved generally, but it is solvable with the help of some new axioms (compare the problem of the continuum hypothesis and the axiom of constructibility). The last, systematicity, refers to the method of arranging the axioms in a systematic manner which enables us to discover new ones.

The last method (that recalls Husserl's "reflective equilibrium" approach to justification) was mentioned by Gödel in the manuscript *The modern development of the foundations of mathematics in the light of philosophy* (1961). Gödel described there in philosophical terms the development of the study of the foundations of mathematics in the 20th century and fitted it into a general scheme of possible philosophical *Weltanschauungen*. Among others, he discussed also Husserl's philosophy, finding in it the method for the clarification of meaning of mathematical concepts.[79] He wrote there (Gödel 1961, p. 385):

> [...] it turns out that in the systematic establishment of the axioms of mathematics, new axioms, which do not follow by formal logic from those previously established, again and again become evident. It is not at all excluded by the negative results mentioned earlier that nevertheless every clearly posed mathematical yes-or-no question is solvable in this way. For it is just this becoming evident of more and more new axioms on the basis of the meaning of the primitive notions that a machine cannot imitate.

Gödel refers here to his famous incompleteness results from (1931). They state that (1) every consistent theory containing the arithmetic of natural numbers contains undecidable propositions and that (2) no such theory can prove its own consistency. Those results showed that neither Hilbert's program of justification

---

[79] This is the only place in which Gödel mentions explicitly Husserl and his philosophy.

of the classical mathematics by means of finitary methods nor Carnap's syntactical program reducing mathematics to its syntax can be realized. Hence the role of mathematical intuition, which can help us to find out deeper meaning and properties of mathematical concepts that are not included in definitions given by axioms. Gödel says in 1961 that there "exists today the beginning of a science which claims to possess a systematic method for such clarification of meaning, and that is the philosophy founded by Husserl". And he continues (Gödel 1961, p. 383):

> Here clarification of meaning consists in concentrating more intensely on the concepts in questions by directing our attention in a certain way, namely, onto our own acts in the use of those concepts, onto our own powers in carrying out those acts, etc. In so doing, one must keep clearly in mind that this philosophy is not a science in the same sense as the other sciences. Rather it is [or in any case should be] a procedure or technique that should produce in us a new state of consciousness in which we describe in detail the basic concepts we use in our thought, or grasp other, hitherto unknown, basic concepts.

This path of Gödel's from the incompleteness results to philosophy is not surprising. In a sense, the incompleteness theorems support and are supported by phenomenological views. They support philosophy because they suggest that an intuition of mathematical essences or a grasp of abstract concepts that cannot be understood on the basis of axioms alone is required in order to solve certain problems and to obtain consistency proofs for formal theories. On the other hand, they are supported by philosophy because the latter gives mathematical essences their due. Gödel claimed that it is necessary to ascend to stronger, more abstract principles and axioms to be able to solve problems from the lower levels (for example, to set theoretic principles to solve number theoretic problems). This idea was strongly supported by the results of Paris, Harrington and Kirby which provided examples of genuine mathematical statements that refer only to natural numbers, that are undecidable in number theory, but that can be solved by using infinite sets of natural numbers.[80]

In the paper *The modern development of the foundations of mathematics in the light of philosophy* (1961), Gödel says also that it is not excluded that every clearly formulated mathematical yes-or-no question can be solved through cultivating our knowledge of abstract concepts, through developing our intuition of essences. In fact, in this way more and more new axioms become evident on the basis of the meaning of the primitive concepts that a machine, i.e. a formal procedure, cannot emulate.

It seems that Gödel settled on Husserl's philosophy because according to it we are directed toward and have access to essences in our experience – and this is a support for Platonism which was Gödel's favourite conception in the philosophy of mathematics.

---

80 Cf. Paris, Harrington (1977) and Kirby, Paris (1982). See also Murawski (1984b).

Conclusion

Husserl's post-psychologistic, transcendental view of mathematics is still a live option in the philosophy of mathematics. As Tieszen writes, it is "compatible with the post-Fregean, post-Hilbertian and post Gödelian situation in the foundations of mathematics" (cf. Tieszen 1994, p. 335). The phenomenological approach to the philosophy of mathematics is still being developed by various authors. The starting point for their considerations are, however, not directly Husserl's works but rather Gödel's considerations. Let us mention here, for example, P. Benacerraf, Ch. Chihara, P. Maddy, M. Steiner, Ch. Parsons and R. Tieszen. They are commenting on Gödel's works concentrating in particular on the problem of mathematical intuition – cf., for example, Maddy (1980), Parsons (1980) or Tieszen (1988).

Mathematical Foundations and Logic in Reborn Poland

Since 1795, there existed no Poland as a sovereign state – in this year the third and the last of the three 18th-century partitions of Poland ended the existence of the Polish-Lithuanian Commonwealth (earlier ones took place, resp., in 1772 and 1790). The partitions were conducted by the Russian Empire, the Kingdom of Prussia and Habsburg Austria, which divided up the Commonwealth lands among themselves progressively in the process of territorial seizures. The cultural situation in each of the occupied parts was different.

The relatively most liberal atmosphere was in the Austro-Hungarian part. There existed two universities: University of Cracow and University of Lvov. The University of Cracow (it acquired its modern name: the Jagiellonian University in 1817) was organized according to the Austrian model, and German was the language of tuition. In the second half of the 19th century, when Austria granted the Poles in Galicia an actual autonomy, crucial changes in the university's situation led to a substantial growth of its scholarly and social significance. Polish was reinstated as the language of tuition. There was a rapid development in both sciences and humanities.

The University of Lvov was founded in 1784, closed in 1805 and reopened in 1817 as a German university. Since the 1870s there took place gradual Polonization of it – Polish was allowed as the language of instruction. Before the First World War, it made its mark especially in the humanities.

The situation in the Russian and Prussian parts was much more difficult. In the Prussian part, there existed no higher education and no university. Prussian authorities were engaged in a campaign of germanization.

In the Russian part, the main scientific centres were Warsaw and Vilnius. The University of Warsaw was created in 1816–1818 – here dominated the humanities. The University in Vilnius (founded by Polish King Stefan Batory in 1579) was then a centre of the sciences. They were both closed after the November Uprising in the 1830s. In 1862 the Main School (Szkoła Główna) was opened in Warsaw – in 1869 it was changed into Russian University.

In 1918 Poland regained its independence after 123 years. The state and its institutions, in particular the whole system of science and education, should be restored. Universities in Cracow, Lvov (since 1919 till 1939 called Jan Kazimierz University in honour of its founder King John II Casimir Vasa) and Vilnius were re-established, the (Polish) university in Warsaw was opened (in fact already in 1915 Germans allowed this in order to win the Poles for their case) and a new university in Poznań (since 1955 called Adam Mickiewicz University) was created. In 1918 was also established a private Catholic Lublin University (Katolicki Uniwersytet Lubelski) in Lublin.

Poland belonged in the interwar period to the leading centres of mathematical logic and the foundations of mathematics in the world. Logical and foundational researches concentrated mainly in Warsaw. One says about Warsaw School of (Mathematical) Logic and Warsaw School of Mathematics, in particular set theory. The former was a part of Lvov–Warsaw School of Philosophy[81] founded by Kazimierz Twardowski (1866–1938), the latter was founded mainly by Zygmunt Janiszewski (1888–1920) and Wacław Sierpiński (1882–1969). One should stress already at the very beginning the close collaboration and mutual influence of logicians (having mainly a philosophical background) and mathematicians in Warsaw. It was possible since Warsaw mathematicians were interested in philosophical problems concerning mathematics; and on the other hand, philosophers and logicians were opened to mathematics and its philosophical and methodological problems.

Let us start by Warsaw School of Mathematics. The story began with the discovery made by Sierpiński. In 1907 he discovered the amazing fact that the plane and the line have the same number of points. Mostowski writes in (1975, p. 9) that when Sierpiński discovered this fact, he wrote to his colleague Tadeusz Banachiewicz, the future professor of astronomy of the Jagiellonian University, who at that time studied in Göttingen, asking him whether this result is known. Banachiewicz answered the question sending a telegram containing the unique word: "Cantor". In this way, he called Sierpiński's attention to Cantor's works – and the latter began to study them. In this way, he learned that this fact had been discovered already 30 years before by Georg Cantor and that it belonged to the fundamental results of the new mathematical discipline, namely to set theory. This was the beginning of Sierpiński's interests in this domain.

Since 1910 he was professor of Jan Kazimierz University in Lvov – he held one of two chairs in mathematics (the other was held by Józef Puzyna). He gave there lectures, among others, in set theory. Add that the opinion – proclaimed sometimes – that Sierpiński's lectures were the first in the world in this new domain of mathematics is erroneous. Earlier lectures in set theory were given by Ernst Zermelo (Göttingen, 1900–1901), Felix Hausdorff (Leipzig, 1901) and Edmund Landau (Berlin, 1902–1903, 1904–1905). Sierpiński wrote also a handbook *Zarys teorii mnogości* [An Outline of Set Theory] (1912).

---

81 On the Lvov–Warsaw School of Philosophy cf. Woleński (1985) and (1989). See also Murawski (2011) and (2014).

At the beginning of the First World War, Sierpiński was sent by Russian authorities to an internment camp in Wiatka – Sierpiński was on vacations when the war began. Thanks to the help of his Russian colleagues, he was able to move to Moscow where he collaborated with Nikolai N. Lusin and learned about the theory of analytic sets which was developed there. In the future, he appeared to be one of the most important persons who developed this new domain of set theory called "descriptive set theory".

In Lvov, Sierpiński aroused young mathematicians' interest in set theory. Among them were Zygmunt Janiszewski, Stefan Mazurkiewicz (1888–1945) and Stanisław Ruziewicz (1889–1941). Janiszewski studied mathematics and philosophy in Zurich, Munich, Göttingen and Paris. In 1911, he took his doctor's degree in the area of topology in Paris, under the supervision of Henri Lebesgue (the examination commission included Henri Poincaré and Maurice Fréchet). In 1913, he presented his *Habilitationsschrift* to the University of Lvov. At the same university in 1913, Mazurkiewicz and Ruziewicz obtained the doctoral degree (under the supervision of Sierpiński) – the former in topology and the latter in the theory of real functions.

When in 1915 the Russian authorities evacuated his university from Warsaw to Rostov upon Don and when some months later Polish university opened in Warsaw, among its first professors were just Z. Janiszewski and S. Mazurkiewicz. At the end of 1918, W. Sierpiński jointed them and received the chair of mathematics. In this way, there were at one place three scholars interested in set theory. Add that Ruziewicz became professor of the Technical University and of Jan Kazimierz University in Lvov as well as a rector of the Academy of Foreign Trade.

Now move to the Warsaw School of Logic. In 1895 Kazimierz Twardowski came to Lvov (from Vienna) and took up the Chair of Philosophy at Lvov University. He gathered around him a group of young scholars who prepared under his supervision their doctoral dissertations. This led to the formation of a group of philosophers having similar philosophical, in particular methodological, background. Some of them were interested in logic, in particular mathematical logic – at that time a new discipline. They founded the so-called Warsaw School of Logic. Main figures were here Jan Łukasiewicz and Stanisław Leśniewski.

Jan Łukasiewicz (1878–1956) studied philosophy under the guidance of Kazimierz Twardowski at the University of Lvov. He obtained his doctor's degree in 1902, and then he continued studies, mainly in Germany and Belgium. In 1906, he presented his *Habilitationsschrift* to the University of Lvov and became a docent of this University. In 1911, he was promoted to an extraordinary professor ('*professor extraordinarius*'). In 1915 he was invited to hold one of the chairs of the newly restored University of Warsaw. In 1920 he was elected to the Chair of Philosophy at the Faculty of Mathematics and Natural Sciences. He was a professor of this university till 1944.

Stanisław Leśniewski (1886–1939) after having graduated from a gymnasium in Irkutsk studied in Germany (among other places in Munich; certain sources speak about his studies in Zurich, Heidelberg and Leipzig). In 1912, he obtained his doctor's degree written under the supervision of Kazimierz Twardowski at the Jan Kazimierz University of Lvov. In 1918 he returned from Moscow to Warsaw where he was given the Chair of Philosophy of Mathematics at the University of Warsaw in 1919. Add that in Moscow he got to know Wacław Sierpiński, who introduced him to the group of Nikolai N. Luzin who dealt with set theory. Łukasiewicz and Leśniewski were in fact co-founders of the Warsaw School of Logic.

One should indicate here one moment characteristic for the tradition of Polish analytic philosophy originated by Twardowski. According to him and his students one should clearly and sharply distinguish world-view and the scientific philosophical work. This idea was particularly stressed by Jan Łukasiewicz. He regarded various philosophical problems pertaining formal sciences as belonging to world-views of mathematicians and logicians, but the work consisting in constructing logical and mathematical systems together with metalogical and metamathematical investigations constituted for him the subject of logic and mathematics as special sciences. Hence philosophical views cannot be a stance for measuring the correctness of formal results. Yet philosophy may serve as a source of logical constructions. One of the consequences of this attitude was the fact that Polish logicians and mathematicians did not attempt to develop a comprehensive philosophy of logic and mathematics (with two exceptions, namely Stanisław Leśniewski and Leon Chwistek). On the other hand, they saw mathematical and philosophical foundations of mathematics as independent although connected in a way and indispensable for understanding logical and mathematical activity. They represented (with two mentioned exceptions) the view guided by the following two principles:

- all commonly accepted mathematical methods should be applied in metamathematical investigations, and
- metamathematical research cannot be limited by any *a priori* accepted philosophical standpoint.

This attitude has its source not only in the tradition going back to Twardowski and his school. Another source can be exemplified by Sierpiński's work on the axiom of choice (AC) and its applications in mathematics. In his French paper (1918) on the role of AC, Sierpiński distinguished two independent questions:

- philosophical controversies around this axiom, and
- its place in proving mathematical theorems.

According to Sierpiński, the second issue should be investigated independently of philosophical inclinations concerning the problem whether the AC is to be accepted or not. This opinion was included in all editions of his textbook on set theory since 1923 (*Zarys teorii mnogości* [An Outline of Set Theory]) to 1965 (*Cardinal and Ordinal Numbers*). In the latter he wrote (p. 95):

> Still, apart from our personal inclination to accept the axiom of choice, we must take into consideration, in any case, its role in the set theory and in the calculus. On the other hand, since the axiom of choice has been questioned by some mathematicians, it is important to know which theorems are proved with its aid and to realize the exact point at which the proof has been based on the axiom of choice; for it has frequently happened that various authors have made use of the axiom of choice in their proofs without being aware of it. And after all, even no one questioned the axiom of choice, it would not be without interest to investigate which proofs are based on it and which theorems are proved without its aid – this, as we know, is also done with regard to other axioms.

This means simply that one should disregard philosophical controversies (and treat them as a "private" matter) and investigate (controversial) axioms as purely mathematical constructions using any fruitful methods.

In 1917 Janiszewski wrote a paper „O potrzebach matematyki w Polsce" [On the needs of mathematics in Poland] (1917). This small (6 pages) paper became a program for the whole generation of Polish mathematicians. Janiszewski postulated there that the future scientific activity should be concentrated on one domain of mathematics. How important this was can be seen from the following anecdote told by E. Marczewski in 1948 (pp. 17–18) where he wrote:

> [...] when [...] in 1911 Puzyna, Sierpiński, Zaremba and Żorawski met in the section of mathematics at the Conference of Scientists and Physicians in Cracow, they found no common subject to discuss: their scientific interests were extremely different.

Janiszewski proposed to erect a new mathematical journal. In 1917 (pp. 15 and 18) he wrote:

> According to the above described project a strictly scientific journal should be erected, a journal that would be devoted exclusively to one of those branches of mathematics in which we have outstanding, really creative and numerous scientific workers. In this journal [...] only papers written in one of the four languages recognized in mathematics as international would be accepted [...]. This journal would contain, beside original papers, also bibliographies of this branch, summaries and even reprints of papers published elsewhere, in particular translations of valuable papers printed in non "international" languages, hence first of all papers written in Polish which are wasted being unknown; finally correspondence: answers to questions [...].
>
> [...] let us come back to the problem of mathematical creativity. Here an appropriate atmosphere can be created only when people will work on common problems. Collaborators are indispensable for a scientist.

> A person working alone does mostly decay. The reasons are not only of a psychic nature (lack of a stimulant): an isolated scientist *knows* much less than those working together. Only the results of researches reach him – they are then already polished and published and this happens usually some years after they have been obtained. An isolated scholar did not see the way in which they have been obtained, he did not experience together with their creators the process of discovering. [...]
>
> Consequently if we do not want to "remain always behind" we should use radical means and to reach the basis of the evil. We should create such a "hotbed" by us! And we can reach this aim only by concentrating most of our mathematicians on the work in one branch of mathematics. It happens now by itself – one should only come to its aid. It is sure that the erection by us of a special journal devoted to one branch of mathematics will attract many [scholars] to the work just in this branch.
>
> The journal would help to create by us such a "hotbed" also in another way: we would become then a technical center of mathematical publications in this branch. Manuscripts of new papers would be sent to us and others would be in contact with us.

A natural candidate for such a branch of mathematics on which the research activity and efforts should be concentrated was just set theory and related domains like topology, theory of real functions, etc. Note that the very term "set theory" did not appear in Janiszewski's paper. It could be the result of the fact that in the same volume a paper by Stanisław Zaremba was published and he was against the "new mathematics". These were just the domains of scientific interests of the group of Warsaw mathematicians who moved to Warsaw from Lvov as well as of a part of Lvov mathematicians. In order to make possible the second of Janiszewski's proposals, a new journal under the title *Fundamenta Mathematicae* was founded. In the first issue of *Fundamenta*, it was written that it is a journal devoted to "set theory and related problems (direct applications of set theory), Analysis Situs [it is called today topology – my remark, R.M.], mathematical logic, axiomatic studies". The first volume appeared in 1920. Unfortunately, Janiszewski did not live to see the publication of this volume – he died at the age of just 31 years on 3rd January 1920 during the influenza pandemic.

Janiszewski and others saw and accepted the connections of set theory with other (both classical as well as being just developed) domains of mathematics and looked at it not as an isolated theory. Add that in the Warsaw School of Mathematics one treated set theory as the foundation of mathematics in the methodological and not in the philosophical (i.e. ontological and epistemological) sense. One treated it rather as an auxiliary (though having the fundamental meaning) theory than as a separate, lonely and self-contained theory. This found its expression in the stress put on its applications in other domains of mathematics. Hence in papers published in *Fundamenta Mathematicae* much more attention was paid to its applications in other theories like topology, function theory or analysis than to its "inner" problems.

One should stress here that the members of the Warsaw School were conscious of the connections between set theory on the one hand and logic and the foundations of mathematics as well as the philosophy of mathematics on the other. This consciousness found its expression among others in the fact that in the Editorial Board of *Fundamenta Mathematicae* beside three mathematicians (i.e. Z. Janiszewski, S. Mazurkiewicz and W. Sierpiński) there were also two logicians mentioned earlier: Stanisław Leśniewski and Jan Łukasiewicz. Both were in the Editorial Board till 1928.[82] Their duty in the board was to take care of the development of mathematical logic and the foundations of mathematics – it was planned that issues of *Fundamenta* will be alternately devoted to set theory and its applications and mathematical logic and the foundations of mathematics. Add that this plan was not fulfilled. The reason was that the number of papers in logic and the foundations submitted to the journal was too small.

The importance attached by the founders of *Fundamenta* to logic and the foundations of mathematics is already stressed by the very name of the journal. This collaboration of mathematicians and logicians (who had a philosophical and not a mathematical background) is really typical for the Warsaw School and distinguishes it from other schools. It should be stressed that they – in particular Łukasiewicz – had extremely good mathematical intuition. Their lectures found very good reception among students of mathematics and were appreciated by them. The collaboration brought many interesting fruits. Thanks to the union of two different traditions and approaches both set theory as well as mathematical logic and the foundations of mathematics were developed in Warsaw in the interwar period in a remarkable way – Warsaw became in fact the first centre of set-theoretical and logical investigations in the world. The collaboration of logicians and mathematicians contributed also to the broadening of the perspective in investigations concerning set theory and the foundations of mathematics. Whereas works by Sierpiński and Mazurkiewicz were devoted to problems connected with paradoxes of the infinity, to the AC and the continuum hypothesis or the descriptive set theory (hence to mathematical problems connected with set theory), works by Leśniewski and Łukasiewicz and their students (Alfred Tarski, Mordechaj Wajsberg, Adolf Lindenbaum, Mojżesz Presburger, Andrzej Mostowski[83] and others) were devoted to metamathematical problems of this theory.

---

82 In 1928, both left the Editorial Board. The reason seems to be certain conflict between Sierpiński and Leśniewski concerning set theory. Sierpiński developed set theory in which sets are understood in a descriptive way, whereas Leśniewski in his mereology developed the theory of sets in a collective sense. Łukasiewicz left the Editorial Board as a sign of solidarity with Leśniewski.

83 In Appendix below, some biographical information on those persons are provided.

* * *

So far we presented the history of Warsaw School of Logic and Warsaw School of Mathematics and their philosophical and methodological background. Let us say at the end some words about scientific achievements of members of those schools.

The most important achievements in logic include works in the classical propositional calculus, non-classical logics, Leśniewski's systems, researches in metamathematics and the semantic conception of truth as well as studies in the history of logic. Researches in the foundations of mathematics were concentrated among others on set theory, model theory, decidability theory or metamathematical studies of axiomatic systems. One should mention here studies on the AC and the continuum hypothesis, studies in the descriptive set theory, infinite combinatorics as well as the idea of determinacy and the concept of measurable cardinal.

Main achievements in logic and the foundations of mathematics can be briefly summarized in the following way:

- Classical propositional calculus:
  - new symbolism, namely parenthesis-free one (Łukasiewicz, Chwistek),
  - construction and study of various axiomatic systems of propositional calculus,
  - from the early 1920s, intensive researches on the metatheory of the propositional calculus and
  - systems of natural deduction (Stanisław Jaśkowski).
- Systems of many-valued logics (Łukasiewicz).
- Leśniewski's systems: Leśniewski did not take part in the before mentioned researches on propositional calculi (with the only exception – the equivalential propositional calculus); his logical studies formed a separate trend. He was rather critical towards mathematical logic or more exactly towards the ways it was cultivated. He criticized Whitehead and Russell's *Principia Mathematica* and proposed new systems called "protothetic, ontology and mereology". Protothetic can be seen as a generalized propositional calculus. Ontology is a systems comprising the calculus of classes, the calculus of relations as well as almost the whole contents of the systems of *Principia*. It is obtained from protothetic by adding to it a functor ε ("is"). Mereology based on protothetic and ontology can be treated as a theory of sets and classes in the collective (mereological) sense.
- History of logic (Łukasiewicz, Jan Salamucha (1903–1944), Innocenty Bocheński (1902–1995), Tadeusz Czeżowski (1889–1981), Zbigniew Jordan (1911–1977) and others).
- Tarski's semantic theory of truth. Theorem on the undefinability of truth. Model theory.

- Metamathematical results: the axiomatic theory of consequence (Tarski; he introduced also the concept of a deductive system), results of Tarski concerning ω-completeness and ω-consistency, his theory of definability, results in decidability theory (e.g. the method of elimination of quantifiers – Tarski, Mojżesz Presburger, Józef Pepis).
- Recursion theory (Kleene-Mostowski hierarchy).
- Computable analysis: Stefan Banach, Stanisław Mazur (1905–1981).
- Studies of set theory and its foundations (Tarski, Lindenbaum, Mostowski): studies of the AC and the continuum hypothesis, descriptive set theory and infinite combinatorics.
- Banach-Tarski theorem on the paradoxical decomposition of the sphere (1924): given a ball in three-dimensional space (without the centre), there exists a decomposition of the ball into a finite number of disjoint subsets, which can then be put back together in a different way to yield two identical copies of the original ball.
- Theory of analytic sets: Sierpiński, Kazimierz Kuratowski (1896–1980).
- Infinite combinatorics – investigations on the abstract problem of measure: Banach, Stanisław Ulam (1909–1984), Tarski, Sierpiński.

Talking about logic and foundations of mathematics in the interwar Poland one should mention also one person who did not belong to described schools, namely Leon Chwistek (1894–1944). His scientific career started in Cracow but was developed in Lvov where he became professor at the Jan Kazimierz University in 1930. His most known achievement is his simplification of Whitehead and Russell's theory of types. He rebuilt the system of Whitehead and Russell in the nominalistic way by constructing a simple theory of types rediscovered later by F.P. Ramsey. In 1924 and 1925, he formulated a pure theory of logical types – a theory of constructive types. In this theory, the nonconstructive objects are rejected but the price for that is the greater formal complication of the system. His logical investigations were – as it was the case by Stanisław Leśniewski – connected with his philosophical ideas concerning logic and mathematics. Moreover, they were in a sense motivated by those ideas. Building semantics he wanted to overcome the philosophical idealism and was against the conception of an absolute truth. He did not content himself with solving particular definite fragmentary problems but – similarly to Leśniewski – attempted to construct a system containing the whole of mathematics.

* * *

So far we presented the picture of Polish logic and the foundations of mathematics between the wars. In the period 1925–1939, Warsaw was the main centre of set-theoretical investigations in the world and one of the best centres of mathematical logic. Let us add at the end one remark. Polish logicians and mathematicians, especially those in Warsaw, were free in their investigations of any philosophical presuppositions. The current trends and views in the philosophy of mathematics, i.e. logicism, intuitionism and formalism, were of course well known (and there appeared papers discussing those tendencies, their meaning and development). But none of them was represented in Warsaw School of Logic and in Polish School of Mathematics. Moreover, they did not represent any other trend; they had no official philosophy of logic and mathematics. This followed from the belief of the autonomy of logic and mathematics with respect to philosophy. Opinions in the field of the philosophy of logic and mathematics were treated as "private" problems, and philosophical declarations were made reluctantly and seldom. If they were made, then it was stressed, directly or indirectly, that these were personal opinions.

This attitude was characterized by Tarski (1954) in the following way:

> As an essential contribution of the Polish school to the development of metamathematics one can regard the fact that from the very beginning it admitted into metamathematical research all fruitful methods, whether finitary or not.

Note that this attitude was in full accordance with the attitude of Polish mathematicians. According to it, one should study the problems using any fruitful methods and making no philosophical presuppositions. There is no need to announce one's philosophical views concerning the investigated problems because this does not belong to scientific duties; this is a "private" affair.

Appendix

ALFRED TARSKI was born Alfred Tajtelbaum in Warsaw on 14 January 1901. In 1923, he changed his surname to "Tarski". In 1918, he began studying biology at the University of Warsaw. Influenced by Leśniewski he abandoned biology and enrolled in mathematics and philosophy. His teachers included Kotarbiński, Leśniewski, Łukasiewicz, Mazurkiewicz and Sierpiński. In 1924, he completed his doctorate under the supervision of Leśniewski. He obtained the *Habilitation* a year later. In 1925–1939, he taught in the Stefan Żeromski Grammar School in Warsaw; and at the same time, he held a temporary position of an associate professor at the University of Warsaw. In August 1939, he left for the USA to participate in the 5th International Congress for the Unity of Science. The outbreak of the war made him stay there. He lived in the USA till the end of his life (in 1946 the University of Warsaw wanted him to accept the post of associate professor but he refused the offer). He lectured at Harvard University (1939–1941), was a visiting professor at City College of New York (1940–1941) and a member of the Institute for Advanced Study in Princeton (1941–1942). He also lectured at the University of California at

Berkeley, where he was appointed as a professor in 1946. There he created a strong research centre for logic and the foundations of mathematics. He died in Berkeley, California, on 26 October 1983.

WAJSBERG MORDECHAJ was born on 10th May 1902 in Łomża, Białostocki district. Studied at the Department of Philosophy of the University of Warsaw, mainly with J. Łukasiewicz. Earned a Ph.D. in 1930. Worked as a teacher in Łomża but the whole time was active as a researcher collaborating with logicians from Warsaw School of Logic. Circumstances, place and date of his death are unknown. His scientific interests and achievements concern axiomatization of two- and many-valued propositional calculus, independence of axioms, modal and intuitionistic logic.

ADOLF LINDENBAUM was born on 12 June 1904 and brought up in Warsaw. He earned a Ph.D. in 1928 under Wacław Sierpiński and habilitated at Warsaw University in 1934. He published works on mathematical logic, set theory, cardinal and ordinal arithmetic, the AC, the continuum hypothesis, theory of functions, measure theory, point set topology, geometry and real analysis. He served as an assistant professor at Warsaw University from 1935 until the outbreak of war in September 1939. He was Alfred Tarski's closest collaborator of the interwar period. Sometime before the middle of August 1941 he and his sister Stefanja were shot to death in Naujoji Vilnia (Nowa Wilejka), 7 km east of Vilnius, by the occupying German forces or Lithuanian collaborators. His most cited works are Lindenbaum's lemma and Lindenbaum algebras.

MOJŻESZ (MOSE) PRESBURGER was born in Warsaw on 27 December 1904. He was a student of Alfred Tarski and is known for, among other things, having invented Presburger arithmetic (arithmetic of addition) as a student in 1929 – he proved the decidability of this theory. Died in the Holocaust, probably 1943.

ANDRZEJ MOSTOWSKI was born in Lvov on 1 November 1913. In 1931 he began mathematical studies at the University of Warsaw. There Alfred Tarski and Adolf Lindenbaum exerted the greatest influence on him. He completed his studies in 1936. In the academic year 1936/1937, he stayed in Vienna and in the year 1937/1938 in Zurich where he listened to the lectures by Kurt Gödel, Hermann Weyl and Wolfgang Pauli. After returning to Warsaw he defended his doctoral dissertation, written under the supervision of Kazimierz Kuratowski (Tarski, who was in fact Mostowski's supervisor, could not fulfil this function since he was not a professor then) in February 1939. Earlier in January 1939, he had begun working for the State Meteorological Institute in Warsaw. During the war he worked as an accountant. In the years 1942–1944, he was engaged in underground teaching at the University of Warsaw. He presented his habilitation thesis to the Jagiellonian University in 1945. From 1946 until his death, he worked at the University of Warsaw: from 1947 as an extraordinary professor and from 1951 – a full professor. He died in Vancouver (Canada) on 22 August 1975.

Note. This text is based on my invited lectures held at the conferences: *École des Mathématiciens Polonais dans l'entre-deux guerres* (Universitè d'Artois, Lens, France, November 2014) and *Mathematical Communities in the Reconstruction after the Great War (1918–1928)* (Centre International de Rencontres Mathématiques, Marseille – Luminy, France, November 2018).

Tarski and his Polish Predecessors on Truth

Co-authored by Jan Woleński

> Almost all researchers who pursue the philosophy of exact sciences in Poland are indirectly or directly the disciples of Twardowski, although his own work could hardly be counted within this domain.
>
> (Tarski 1992, p. 20)

This is Tarski's description of the genesis of Polish investigations in mathematical logic, or more precisely those done inside the Lvov–Warsaw School.[84] Since the semantic theory of truth belongs to the philosophy of the exact sciences, we conclude that Tarski considered himself as a member of Twardowski's heritage.[85] Sociologically it is obvious that he was - in fact Tarski was a student of Kotarbiński, Leśniewski, and Łukasiewicz, i.e. direct disciples of Twardowski. Not very much is known, however, about direct contacts between Tarski and Twardowski. Almost all the information we have comes from Twardowski's *Diary*.[86] On 7 September 1927, Twardowski described Banach's lecture on the concept of limit at the first Polish Mathematical Congress and says that "there came several of my acquaintances from Warsaw, except Łukasiewicz, Sierpiński, Tarski and others" – a remark that at least let us know that Twardowski took Tarski to be among his acquaintances.

Perhaps the most interesting record in the *Diary*, however, concerns Tarski's chapter on truth, delivered in Polish Philosophical Society in Lvov on 15 December 1930: "Very interesting and also very well construed". Other fragments of Twardowski's *Diaries* about Tarski mention the problem of the latter's candidacy for a professorship in Lvov (Twardowski supported Tarski; see also Feferman and Feferman 2004, pp. 66–688), mutual meetings (Tarski often visited Twardowski in Lvov), exchanges of letters and the preparation of the German version of Tarski (1933) for *Studia Philosophica* (it was published in 1936). We have also a letter of Twardowski to Leśniewski (see Feferman and Feferman 2004, pp. 100–102) written in 1935 in which the former supports Tarski's professorship in Warsaw. Although

---

84 See Skolimowski (1967) and Woleński (1989) for detailed presentations of this philosophical formation.
85 Currently there is a problem with spelling the name "Lvov". "Lwów" is the Polish version, "Lviv" – Ukrainian. Some Ukrainians say that "Lvov" is a Russian word. We take the last as the English spelling of "Lwów".
86 See Twardowski (1997), Part I, p. 323, Part II, pp. 110–113, 176, 179, 180, 205, 296, 331, 336, 352, 369, 372.

the relations between Twardowski and Tarski had never been particularly close, all accessible evidence allows one to assert that they were good.

Here, however, we are much more interested in the substantial influence of Twardowski and his direct students on Tarski's work on truth than in obvious sociological links. We intend to show that this influence was important. Although the mathematical side of the semantic theory of truth is independent of its philosophical background, the latter cannot be properly understood without taking into account the aletheiology (we propose this word as an equivalent for "philosophy of truth") developed by Tarski's Polish philosophical ancestors. We will discuss the views of Twardowski, Łukasiewicz, Leśniewski, Zawirski, Czeżowski and Kotarbiński.[87] The last philosopher will be treated more extensively than the rest, because his influence on Tarski was greater than that of anybody else, save perhaps Leśniewski. However, as Leśniewski's aletheiology is extensively treated by Arianna Betti (2008), we restrict our remarks on Leśniewski to a very few.[88]

Twardowski

Twardowski's main work on truth (1900) concerned the problem of aletheiological relativism. His understanding of truth and its absoluteness or relativity is as follows (1900, p. 148):[89]

> The term "a truth" designates a true judgement. Therefore, all judgements that are true, that possess the characteristic of truthfulness, are truths. Hence, it is always possible to use the expression "a true judgement" instead of the term "a truth". It then follows that expressions "relative truth" and "absolute truth" mean the same as the expressions "relatively true judgement" and "absolutely true judgement."
>
> Those judgements that are unconditionally true, without any reservations, irrespective of any circumstances, are called "absolute truths" – judgements, therefore that are true always and everywhere. On the other hand those judgements that are true only under certain conditions, with some measure of reservation, owing to particular circumstances, are called "relative truths", such judgements are therefore not true always and everywhere.

Twardowski, following Bolzano and Brentano, rejected the view that there exist relative truths, though he was interested in categorizing the reasons some had for accepting some truths as relative. One such reason, according to Twardowki's

---

87 Although we concentrate on aletheiology, our considerations on nominalism – cf. below – go beyond the problem of truth. Nominalism was one of the most intriguing of Tarski's views.

88 This chapter uses some material published earlier in Woleński and Simons (1989), Woleński (1990, 1993a, 1993b, 1994b, 1995, 1999); see also Vuissoz (1998).

89 If our bibliography lists a translation or another edition of an original work, page references are to later sources.

survey, stems from elliptical formulations of some judgements through the use of occasional words, like "now", "here", 'I', etc. – for example, the apparent relativity of the truth of "It is raining today" to a time and a place. Other relativist arguments, he notes, point out the relativity of various evaluations (for example, of "bathing is healthy") to some salient person, or appeal to the view that empirical hypotheses are neither true or false, but always only probable. Twardowski held that all these arguments are erroneous. In particular, on his view one should sharply distinguish sentences from complete propositions. Only the former can appear as relatively true or false. For example, the sentence "Today it is raining in Lvov" does not express a complete proposition. After eliminating "today" and inserting a concrete date, we obtain a sentence that does express a complete proposition, for example, "December 17, 1899 it is raining in Lvov", which is absolutely true or false. The same treatment applies to evaluations, because we should complete "bathing is healthy" by indicating a person. Hence, though some sentences are relatively true or false, only complete propositions are absolutely true or false. Twardowski also pointed out that the relativity of truth is at odds with principles of excluded middle and non-contradiction.

If we analyse the most typical case of aletheiological relativity, that with respect to time, the view that truth is absolute can be displayed by two sub-theses:

(1) A proposition A is true at t if and only if it is true at every $t' \leq t$, and
(2) A proposition A is true at t if and only if it is true at every $t' \geq t$.

The first sub-thesis expresses the principle of sempiternality of truth (A is true at t if and only if it is true at every earlier moment), while the second gives the principle of the eternality of truth (A is true if and only if it is true at every later moment). If (1) and (2) are accepted, truth does not need to be indexed by time. Twardowski, in accord with his distaste for relativism, did in fact accept (1) and (2) and thereby held that truths are, if ever true, always true.

Turning to his other views on truth, Twardowski had some reservations about the concept of correspondence. In this he followed Brentano. Twardowski's own definition of truth was as follows (1975, p. 208):

(3) An affirmative judgement is true if its object exists, a negative judgement, if its object does not exist. An affirmative judgement is false if its object does not exist; a negative judgement, if its object exists.

Twardowski considered (3) to be a version of Aristotle's definition given in *Metaphysics* 1011b (to say of what is that it is not, or of what is not that is, is false, while to say of what is that it is, or of what is not that is not, is true). On the other hand, Twardowski rejected another of Aristotle's formulations, namely that of *Metaphysics* 1051e, which defines truth in terms of thinking the separated to be separated and the combined to be combined. The main argument Twardowski accepted as telling against this second Aristotelian definition is that it is inconsistent with the "idiogenic" account of judgements as *sui generis* acts, rather than combinations of presentations: as truths were unitary on Twardowski's view, truth could not be defined in terms of combination and separation as Aristotle's 1051e has it. Twardowski directed this same sort of objection against Russell, arguing further the Russellian notion of a fact was unclear (Twardowski 1975). In general, he had doubts as to whether typical wordings of the correspondence theory (the theory of "transcendent correspondence" as he called it) were satisfactory. He accused them of being based on unclear metaphysical assumptions concerning what propositions were. Although he agreed that correspondence theories do not offer criteria of truth which would allow one to recognize which judgements were true, he did not consider this sufficient ground for objection. Twardowski also criticized various non-classical definitions of truth, and in particular he argued against pragmatism and coherentism on the grounds that they violated the metalogical principles of excluded middle and contradiction.

To sum up, many of Twardowski's views became important for the further development of thinking about truth in Poland.[90] First of all, his defence of the absolute concept of truth was accepted by most Polish philosophers, an aspect of thought about truth that became important for the discussions of many-valued logic. Here Twardowski and Leśniewski defended the view that there are only two truth values. By contrast, Kotarbiński at first admitted judgements which are indefinite, at least until Leśniewski convinced him to accept strict bivalence. On the other hand, as is well known, Łukasiewicz agreed with eternality, but rejected sempiternality as leading to fatalism. Secondly, Twardowski was the first to point out that some metalogical laws (of excluded middle, of non-contradiction) are associated with the absolute character of truth. Thirdly, Twardowski's criticism of non-classical truth-definitions (for instance, of the "utilitarian" conception) became standard in Poland. Fourthly, his doubts concerning the usual formulation of the correspondence theory were shared by his students. As we will see later in this chapter, all of these views find expression in Tarski's work.

Finally, although it was not directly related to the problem of truth, Twardowski introduced (see Twardowski 1912) a distinction between actions and products which was applied by Polish philosophers to the analysis of all mental activities, including the use of language. In particular, this distinction allowed a fruitful approach to the meaning of linguistic expressions. A special group of acts, the psycho-physical, included linguistic activities. Every psycho-physical act has its

---

90 See Woleński (1989) for a full account of the development summarized here with bibliographical references.

content, which is intuitively apprehended and objectivized as the meaning of a given expression. Moreover, every act has its object, i.e. an entity to which the act is directed.[91] These views will be directly relevant to our discussion of Tarski's somewhat fraught remarks on the bearers of truth.

Łukasiewicz

Although Łukasiewicz is famous for his contributions to many-valued logic, we will omit all problems related to this topic, as Tarski was not particularly interested in it. Although he did some technical work in the area in the 1920s and 1930s, he had little respect for this line of logical investigations in the later stages of his career. He wrote (Tarski 2000, p. 25):

> [...] I hope that no creators of many-valued logics are present, so [...] I can speak freely – I should say that the only one of these systems for which there is any hope of survival is that of Birkhoff and von Neumann. [...] This system will survive because it does fulfill a real need.

(The quotation is likely an allusion to Łukasiewicz and signals rather poor relations among them after the Second World War.) However, there are other points in Łukasiewicz's views on truth which are important in the present context. Although Łukasiewicz considers propositions as proper bearers of truth, he locates them as existing in a language (see Łukasiewicz 1910, *passim*). Hence, Twardowski's distinction between sentences and propositions became of secondary importance to Łukasiewicz. In his later works, he always regarded sentences as the objects of logical investigation. Łukasiewicz, following Twardowski, sharply distinguished truth and its criteria (see Łukasiewicz 1911). He proposed the following definition as a version of Aristotle's from *Metaphysics* 1011b (see Łukasiewicz 1910, p. 15):

(4) An affirmative proposition is true if it ascribes a property to an object, which is possessed by this object; a negative proposition is true if it rejects a property, which is not possessed by a given object.

Łukasiewicz also gave a version of the Liar paradox which was used by Tarski (see 1915; it is unfortunate that the relevant passages of this chapter are not included into Łukasiewicz 1970). It was as follows:

(5) The sentence printed in the line m on the page n of this book is false,

---

91 The same concerns purely mental acts, i.e. acts without physical components: they, too, had entities toward which they were directed. However, most of Twardowski's students considered thinking as essentially linked with the use of language. Thus, the distinction between purely mental and psycho-physical acts was widely rejected in the Lvov–Warsaw School.

where m and n refer to the appropriate line of the appropriate book. Łukasiewicz's response was to maintain that (5) is ill-formed and as such cannot be a value of a propositional variable.

Łukasiewicz also worked on the foundations of probability (see Łukasiewicz 1913). In particular, he argued that sentences are true or false and, thereby, cannot be considered as merely probable. Probability can be ascribed only to indefinite sentences, i.e. formulas with free variables. Now if Px is such a formula, $p(Px)$ (= the probability of Px) is its logical value, which is measured by the relation of the number of values satisfying Px to the number of all possible values. In a particular case, Px is true if all values satisfy it, and false if it is satisfied by no value. We can say that truth defined in such a way conforms to the following condition:

(6) Px is true if and only if $\forall x Px$ is true.

The relation of this to certain aspects of Tarski's treatment of truth will be discussed below.

Zawirski, Czeżowski

Although as of the present writing no definitive historical link can be established between Zawirski and Czeżowski and Tarski, the contributions of these two authors clearly anticipate Tarski's work and they bear mention here. Zawirski (1914), like Łukasiewicz, construed propositions as items of a language and denied that truth and falsehood could have degrees. Following Twardowski, he favoured the idiogenic theory of judgements and Aristotle's formulation from *Metaphysics* 1011b. However, he defended (1914, pp. 57–58) the nihilistic account of truth, saying that every attribution of truth or falsehood is either an assertion or denial of that to which truth is apparently attributed, more precisely, the assertion of reality or the rejection of reality. This account of truth was also discussed by Kotarbiński, as we will see below.

Czeżowski (see Czeżowski 1918, p. 7) was the first author in Poland to focus on the formula later called the T-scheme:

> Truth is an characteristic attribute of sentences [note 'sentences', not 'propositions' – our remark, R.M., J.W.], [...]. We assert truth or falsehood about every sentence. However, truth is a property of a particular importance. If a certain sentence A is true, the sentence *A is true* is also true, if one of them is false, the same simultaneously concerns the second: the sentences A and *A is true* are equivalent.

One should perhaps add that the equivalence of A and *A is true* occurs in Couturat (1905, p. 84), translated into Polish in 1918.

Kotarbiński

As Leśniewski (and Łukasiewicz) were his masters in logic, so Kotarbiński was in philosophy. Tarski's main background was in mathematics but he very seriously studied philosophy under Kotarbiński. Tarski really revered Kotarbiński. One of the indications of this can be seen in the dedication of his 1956 collection (Tarski 1956) of fundamental papers *Logic, Semantics, Metamathematics*. The dedication reads: "To his teacher TADEUSZ KOTARBIŃSKI. The author". The dedication for the second edition (1983) which appeared after the death of Kotarbiński was: "To the memory of his teacher TADEUSZ KOTARBIŃSKI. The author". This is remarkable when one takes into account that Tarski had many teachers who influenced his scientific interests, in particular Łukasiewicz, Sierpiński and Leśniewski, the last of whom was his dissertation advisor. When asked by doctoral students in Berkeley who his teacher was, Tarski replied "Kotarbiński". Leśniewski's name was never mentioned. Add also that Kotarbiński's photo had a privileged position on Tarski's desk. People who observed meetings of Kotarbiński with Tarski were very impressed by their mutual relations and the great respect of the pupil for his teacher.[92] He translated (together with David Rynin) into English Kotarbiński's chapter „Zasadnicze myśli pansomatyzmu" [The Fundamental Ideas of Pansomatism]. The chapter was originally published in Polish in 1935, the translation appeared in *Mind* in 1955 and has been also included into Tarski's *Collected Works* (1986, vol. 3) on the explicit request of Tarski himself.[93]

Kotarbiński's doctrine of reism (called also pansomatism or concretism) is a form of physicalistic nominalism. The main reistic thesis is that there exist only singular, spatio-temporal, material things, some of them equipped with psyche. Thus there are no abstract entities like properties, relations or state of affairs.[94] Kotarbiński was very strongly influenced by Leśniewski's logical and philosophical ideas. Leśniewski was also a nominalist. His calculus of names, called ontology (LO, for brevity), was considered as the logical basis of reism. The concept of an object as defined in LO became the central tool for Kotarbiński. According to LO, a is an object if and only if a is something; a exists if and only if something is a. One can prove that LO implies that only individual objects exist. Thus, things in the reistic

---

92 Marian Przełęcki told us about the meeting in Bucharest at the International Congress of Logic, Methodology and Philosophy of Science in 1971, in which it was clear that there was a great affection between the two men.
93 Kotarbiński was the first reviewer of Tarski (1933) (see Kotarbiński 1934).
94 Since we are not interested in reism as such, we do not enter into a more detailed analysis of this view. For assessments of reism, sympathetic as well as critical, see the papers collected in Woleński (1990b). Let us note that reism was accepted by Brentano in his later philosophy. See Woleński (1996) for comparisons of various forms of reism.

sense are individuals as defined in LO, and being material is their additional universal attribute. Things are usually mereological complexes, i.e. aggregates of material pieces. This idea was formally elaborated in Leśniewski's mereology.

Reism determined some essential features of Kotarbiński's theory of truth (see Hiż 1966, Woleński 1990a for a general account). Since from the point of view of reism there are no propositions (they are abstract objects and rejected by reism), the predicate "is true" cannot be applied to such entities. Although Kotarbiński did not admit propositions in the psychological sense either (because he also banished abstract contents from the furniture of the world), he recognized the existence of subjects performing mental acts, that is, if somatic bodies with mental acts as their proper parts. Hence, as in the case of later Brentano, truth can be attributed to acts of thinking or speaking of concrete persons, e.g. one can think or speak truly or not. This use of "truly" indicates that Kotarbiński to some extent advocated a kind of adverbial theory of truth (see Pasquerella 1989). However, for Kotarbiński, sentences understood as inscriptions or sounds are the principal bearers of truth on the reistic position. Although he noticed that "is true" is predicated both of acts (thoughts) as well as of sentences, and considered this situation to be puzzling, in the end he agreed that it was tolerable. Kotarbiński distinguished at least three interpretations of sentences (*Elementy*, pp. 104–105): idealistic (sentences are ideal objects), psychologistic (sentences are psychical entities) and nominalistic. He adopted the last. For Kotarbiński, inscriptions or sounds are things in the normal sense.[95] This is clearly expressed in *Elementy*, his *opus magnum* (1929, p. 104 and p. 109, resp.):[96]

> [...] in the nominalistic (outward) interpretation, *propositio* [that is, a sentence – our remark, R.M., J.W.] [...] means [...] the symbol itself, the inscription, the statement, the linguistic phrase or formulation.

> [...] There are no "truths" or "falsehoods", if they should be any so-called "ideal objects", some so-called "objects from the world of content." There are only persons who are thinking in a true way and persons thinking in a false way as well as true sentences and false sentences. Hence terms "truth" and "falsehood" will be proper names, and they will be non-empty, if by "truth" one will understand "true sentence" and by "falsehood" – "false sentence."[97]

---

95 See Rojszczak (2005) for a detailed account of truth-bearers, including Brentano and other adverbialists.
96 Since the title of the English translation (Kotarbiński 1966) of his 1929 is very unfortunate we will give – when referring to this work – the first word of the Polish title. However, we quote after the English edition.
97 The second passage supplements the first one, but also contains a combination of adverbialism and reism.

Kotarbiński was an advocate of the absolute character of truth and an opponent of the relativist approach; he closely followed Twardowski in this (observe also the similarity between "truth" and "true sentence" in both philosophers), although the position of Kotarbiński was perhaps slightly weaker than that of his teacher, perhaps because Kotarbiński in this early stage had some reservations concerning the absoluteness of sentences about the future. (See Woleński 1990 on this problem.) In Kotarbiński (1926, p. 135) he remarked:

> The controversy between absolutism and relativism has not been sufficiently explained so far [...], but at least in the domain of scientific sentences absolutism is undoubtedly right.

According to Kotarbiński, being true or false does not depend on who is uttering the given sentence or on the circumstances in which they do so. In *Elementy* he wrote (p. 113):

> The reader has certainly seen that the position of the relativism is weaker. Hence, though relativism attracts some minds today (see, e.g., writings of pragmatists) as it did in the period of Greek sophists [...], so among good specialists in the domain of logic relativism is not popular.

Kotarbiński distinguished the real from the verbal understanding of truth.[98] This seems to be his original contribution to the theory of truth.[99] According to him, in some contexts the predicate "true" (resp. "false") is not necessary; it plays exclusively the role of a stylistic ornament and does not add anything to the content of a sentence. One can reformulate such a sentence without using the term "true" (or "false"). Hence the statement "The proposition that Warsaw is the capital of Poland is true" can be replaced by the statement "Warsaw is the capital of Poland". In this use, "true" conforms to the "nihilistic" theory of Zawirski.

But Kotarbiński notices that this is not always the case. For example, the expressions "The theory of relativity is true" or "What has been said by Plato is true" cannot be reformulated in this way. By omitting the predicate "is true" in these sentences one gets expressions not only of other senses but even of a different grammatical type, namely they become names and not sentences. Hence in various contexts the predicate "is true" ("false") is necessary and cannot be eliminated. In such cases the adjectives "true" and "false" are used in a real, and not merely verbal, sense. The nihilist account of truth in Kotarbiński's sense corresponds to some extent to a variety of views covering the redundancy theory (Ramsey 1927), deflationism (Field 1994), minimalism (Horwich 1999), prosentialism (Grover

---

98 In Kotarbiński (1926) the terms "real" and "nihilistic" were used. See also Kotarbiński (1934).
99 Unfortunately, Kotarbiński did not point out representatives of these views. So we do not know whether he had Zawirski (see above) in his mind. Brentano could be another possibility, because he anticipated the prosentential account of truth.

1992) or disquotationalism (Quine 2004). According to nihilism as Kotarbiński understood it, the sentence "Snow is white is true" (or "It is true that snow is white") says no more than does the sentence "Snow is white". Hence nothing is added to the sentence by adding the suffix is true. Hence one can claim that the predicate is true is empty and adds nothing. So it does not represent or attribute any particular property to its subject. The fact that in our language there are the predicates "is true" and "is false" is of a historical but not of a logical interest. As has been said above, Kotarbiński accepted the nihilistic theory of truth only with respect to verbal (in fact: redundant) uses of the predicate "is true" ("false") and claimed that those predicates are indispensable in various important contexts. Hence the nihilistic theory does not suffice.

Kotarbiński, in *Elementy* (Chapter 3, §17), understood the classical and utilitarian conceptions of truth as the two basic conceptions. According to the first a truth is that which corresponds to or is in agreement with reality, and according to the second, "true" means "useful" (in some respect). One of the forms of utilitarian understanding is pragmatism, which claims that truth is just the property of a proposition which makes an action based on it efficient. Having distinguished those two senses' Kotarbiński explicitly expressed his preference for the classical understanding. On the other hand, he was aware that the phrase "accordance with reality" is not precise enough and has a rather metaphorical character when understood by analogy to pictures or copies. In *Elementy* he wrote (pp. 106–107):

> Let us [...] pass to the classical doctrine and ask what is understood by "accordance with reality". The point is not that a true thought should be a copy or simile of the thing of which we are thinking, as a painted copy or a photograph is. A brief reflection suffices to recognize the metaphorical nature of such comparison. A different interpretation of "accordance with reality" is required. We shall confine ourselves to the following: "John thinks truly if and only if John thinks that things are so and so, and things are in fact so and so."

As we see, Kotarbiński preferred here unequivocally weak over strong correspondence, i.e. he did not invoke such notions as simililarity or isomorphism in order to explain the concept of correspondence (see Woleński 1993b for a more detailed account of the distinction of strong and weak correspondence).

Tarski's views related to the previous sections

Tarski considered his analysis of the concept of truth as a logical (mathematical) enterprise, as well as a philosophical one, as is explicitly asserted in his treatise on truth (1933) in both its opening and closing passages (pp. 152 and 266–267, resp.):

> The present article is almost wholly devoted to a single problem – the definition of truth. Its task is to construct – with reference to a given language – a materially adequate and

formally correct definition of the term 'true sentence'. This problem [...] belongs to the classical questions of philosophy [...].
[...] in its essential parts the present work deviates from the main stream of methodological study [that is, metalogical or metamathematical; the scope of methodological study should be seen here in a wider sense than in the Hilbert school, that is, as not restricted to finitary proof theory – our remark, R.M., J.W.]. Its central problem – the construction of the definition of true sentence and establishing the scientific foundations of the theory of truth – belongs to the theory of knowledge and forms one of the chief problems of philosophy. I therefore hope that this work will interest the student of the theory of knowledge and that he will be able to analyse the results contained in it critically and to judge their value for further research in this field, without allowing himself to be discouraged by the apparatus of concepts and methods used here, which in places have been difficult and have not been used in the field in which he works.

Hence, it is quite legitimate to look at Tarski's philosophical background. As far as the matter concerns terminology and a broad philosophical perspective, Tarski usually refers to *Elementy*:

A good analysis of various intuitive conceptions concerning the notion of truth is contained in Kotarbinski's book [*Elementy*]. (1932, p. 615)
[...] in writing the present article I have repeatedly consulted [*Elementy*] and in many points adhered to the terminology there suggested. (1933, p. 153, note 1)
A critical discussion of various conceptions of truth can be found in [*Elementy*]. (1944, p. 695, note 6)

Yet the substantial links between Tarski and other philosophers from the Lvov–Warsaw School are at least as important. We address these issues in this section. A special problem will be discussed later on.

Classical, correspondence, etc.

Tarski followed Kotarbiński in understanding the contrast between the classical and utilitarian truth-definitions as the main opposition in aletheiology.[100] He also (see Tarski 1944, p. 698, note 38) referred to Kotarbiński as a person who interpreted the semantic conception of truth as a version of the classical theory. Thus, Tarski's claim that he semantically developed the classical tradition was entirely coherent with Twardowski and his tradition.[101] In fact, Tarski adhered to the classical

---

100 Note, however, that other philosophers from the Lvov–Warsaw School, in particular, Leśniewski, considered aletheiological pragmatism as the most important rival of the classical position.
101 This does not mean that every philosopher from this school accepted the classical definition, but the exceptions were rare, e.g. the consensus account was advocated by Poznański and Wundheiler.

correspondence conception of truth and that in just the formulation given by Kotarbiński. At the very beginning of Tarski (1933) we read (p. 155):[102]

> [A] true sentence is one which says that the state of affairs is so and so, and the state of affairs indeed is so and so.

An important note is associated with this passage, which reads:

> Very similar formulations are found in Kotarbiński [1929] [...] where they are treated as commentaries which explain approximately the classical view of truth.

In several places Tarski stressed that his conception of truth coincides with the intuitive classical Aristotelian one and refers to various authors, a fact stressed by reviews such as Kotarbiński (1934) and Scholz (1937). Commenting on intuitions underlying the semantic definition of truth, he wrote (1944, pp. 666–667):

> We should like our definition to do justice to the intuitions which adhere to the *classical Aristotelian conception of truth* – intuitions which find their expression in the well-known words of Aristotle's *Metaphysics*:
> *To say of what is that it is not, or of what is it not that it is, is false while to say of what is that is, or of what is not that it is not, is true.*
> If we wished to adapt ourselves to modern philosophical terminology, we could perhaps to express this conception by means of the familiar formula:
> *The truth of a sentence consists in its agreement with (or correspondence to) reality.*
> (For a theory of truth which is to be based on upon the latter formulation the term "correspondence theory" has been suggested).
> If, on the other hand, we should decide to extend the popular usage of the term "designate" by applying it not only to names, but also to sentences, and if we agreed to speak of the designata of sentences as "states of affairs," we could possibly to use for the same purpose the following phrase:
> *A sentence is true if it designates an existing state of affairs.*
> However, all these formulations can lead to various misunderstandings, for none of them is sufficiently precise and clear (though this applies much less to the original Aristotelian formulation than to either of the others); at any rate, none of them can be considered a satisfactory definition of truth.

Three points are worthy of note. Firstly, Tarski, like Twardowski, Łukasiewicz and Kotarbiński, took the quoted passage from Aristotle's *Metaphysics* as the best approximation of the Stagirite intuition about truth. Secondly, the term "state of affairs" has here a misleading technical ontological connotation (see footnote 102).

---

102 A caution is required here. In particular, the phrase "state of affairs" has no technical meaning, i.e. it does not commit us to an ontology of states of affairs. Tarski (or rather Woodger, the translator of Tarski 1933 into English) used it as a substitute for Kotarbiński's (see below) "things are so and so".

Thirdly, and most importantly in the present context, Tarski considers formulations with "agreement" or "designating states of affairs" as not quite satisfactory. This appears to be a legacy of Twardowski's and Kotarbiński's scepticism concerning the concept of correspondence.

The Liar paradox, satisfaction and the T-scheme

Tarski (1933, p. 157) explicitly used Łukasiewicz's version of the Liar paradox (see above), but he never said that self-referential sentences are ill-formed. Tarski's view was rather that they should not appear in properly constructed formalized languages.[103] In this sense, he would agree with Łukasiewicz that the Liar sentences and other self-referential constructions could not be values of sentential variables. The defective character of such sentences consists in their role in generating semantic antinomies. In general, we can say that according to the logicians of the Lvov–Warsaw School, good symbolic notations should not lead to contradictions caused by rules of formation.

Tarski defined truth (he also identified truths with true sentences) via satisfaction: a sentence is true if and only if it is satisfied by all sequences of objects. It is difficult to say whether Tarski was influenced by Łukasiewicz in this respect (see above), although Łukasiewicz (1913) was among the best-known philosophical papers in Poland. Anyway, (6) is a consequence of Łukasiewicz's account as well as the semantic definition of truth. According to Tarski, the intuitive content of this definition is captured by the T-scheme "A is true if and only if A" (with additional constraints concerning protection against antinomies), but it is not clear whether Tarski recognized that Czeżowski (see above) formulated the equivalence of A and "A is true" as the rule governing the concept of truth.

Absolutism vs. relativism

In Tarski (1933, pp. 199–200), we find the distinction between the absolute concept of truth and that expressed by the phrase "true sentence in an individual domain of individuals". According to Tarski, the former is a special case of the latter. It is unclear whether Tarski himself attributed any philosophical significance to this distinction. On the other hand, the semantic definition of truth was used (see Kokoszyńska 1936, 1948, 1951) for making precise the distinction in question. In particular, Kokoszyńska, who was a good expert on Tarski's views, considered his theory as absolute and argued that the reference to models or languages does not entail (see also Woleński 1994) the relativity of truth. Although we have no explicit

---

103 Concerning the Liar paradox, Tarski was much more influenced by Leśniewski, but we omit this issue.

comment from Tarski about this issue, we can say that his ideas are coherent with absolutism.[104]

The verbal and real use of "is true"

Tarski in various places referred to Kotarbinski's distinction between real and verbal usage of the predicate "is true" and to the nihilistic theory of truth (see Tarski 1944, 1969).[105] In particular, Tarski shared Kotarbinski's opinion that the predicate "is true" is not always eliminable. He generalized Kotarbinski's argument in a very interesting way (Tarski 1944, pp. 682–683):

> Some people have [...] urged that the term "true" in the semantic sense can always be eliminated, and that for this reason the semantic conception of truth is altogether sterile and useless. And since the same considerations apply to other semantic notions, the conclusion has been drawn that semantics as a whole is purely a verbal game and at best only a harmless hobby.
> But the matter is not quite simple. [...] The sort of elimination here discussed cannot always be made. It cannot be done in the case of universal statements which express the fact that all sentences of a certain type are true, or that all true sentences have a certain property. For example, we can prove in the theory of truth the following statement:
> *All consequences of true sentences are true.*
> However, we cannot get rid here of the word "true" in the simple matter contemplated.
> Again, even in the case of particular sentences having the form "*X is true*" such a simple elimination cannot always be made. In fact, the elimination is possible only in those cases in which the name of the sentence which is said to be true occurs in a form that enables us to reconstruct the sentence itself.

The non-eliminability of "is true" was important for Tarski, because it armed him against the view that "the semantic conception of truth is altogether sterile and useless."[106]

---

104 Note, however, that Jan Tarski, the son of Alfred, told one of us (Jan Woleński) that his father considered the absoluteness of truth as truth's important feature.

105 In Tarski (1969) we find an explicit reference to Kotarbiński and the assertion that the name "nihilistic theory of truth" was suggested by him.

106 Although we agree with Tarski, we would not like to suggest that this issue is uncontroversial. The sentence "all consequences of true sentences are true" can be rendered in the context of the redundancy theory as follows:

$$\forall A \forall B[(A \in Cn(B) \wedge B) \to A]$$

Applying the T-scheme gives: if B is true, so is A. However, this translation is much more complicated than the original and assumes a quite considerable amount of logic, for instance, the rules for quantifiers for propositional variables. It is also debatable whether the fundamental limitative theorems hold without a precise truth-definition, but we do not enter into this topic.

Truth-bearers

Tarski, similarly to Kotarbiński, Leśniewski and Łukasiewicz, claimed that the predicate "is true" (resp. "is false") should be applied only to sentences. He did not exclude other bearers of truth, such as thoughts or judgements, but his nominalistic preferences (inherited from Leśniewski and Kotarbiński) determined that he considered linguistic expressions, in particular, sentences as primary bearers of semantic properties.[107] This view, together with the role of items of the sentential syntactic category in logic, led him to construe a language as a set of sentences.[108] Hence, a closer analysis of the concept of a sentence was of the utmost importance, for on Tarski's view the issue was relevant to nominalism. He wrote (1930, p. 62):

> Sentences are most conveniently regarded as inscriptions, and thus as concrete physical bodies.

According to this explanation, language consists of expressions conceived as tokens. Yet Tarski was fully aware of the fact that this purely nominalistic theory of language created serious difficulties for logic, particularly metalogic and metamathematics. This led him to the idea that linguistic expressions should be considered not as concrete inscriptions but as types, i.e. as shapes of tokens (mathematically speaking, types are classes of abstractions from similar tokens). Tarski expressed this new approach in the following way (he refers to Kotarbiński; also to *Principia Mathematica* of Whitehead and Russell) (1933, p. 156):

> Statements (sentences) are always treated here as a particular kind of expression, and thus as linguistic entities. Nevertheless, when the terms 'expression', 'statement', etc., are interpreted as names of concrete series or printed signs, various formulations which occur in this work do not appear to be quite correct, and give the appearance of a widespread error which consists in identifying expressions of like shape. [...] In order to avoid both objections of this kind and also the introduction of superfluous complications into discussion, which would be connected among other things with the necessity of using the concept of likeness of shape, it is convenient to stipulate that terms like 'word', 'expression', 'sentence', etc., do not denote concrete series of signs but the whole class of such series which are of like shape with the series given.

---

107 See below for a more detailed account of Tarski's nominalism. Let us add a word about Łukasiewicz in this context. He accepted nominalism with respect to truth-bearers more as a useful practical solution than a theoretically justified standpoint (see also footnote 109 below).

108 Adopting this view in logic consisted in a radical departure from the traditional logic for which sentences (or propositions or judgements) and concepts (notions, names) constituted equally important building-blocks of logic.

Tarski considered this new account as more convenient for logic.[109]

It is interesting that Tarski carried the analysis of the concept of a sentence beyond that of his predecessors. In Tarski (1933) we find at least four different understandings of this concept:

(a) an expression of a special syntactical category (this interpretation is the most suitable for formalized languages):

> "Among all possible expressions which can be formed with these signs those called sentences are distinguished by means of purely structural properties." (p. 166)

(b) a sentential function of a kind (also good for formalized languages):

> "x is a sentence (or a meaningful sentence) – in symbols $x \in S$ – if and only if x is a sentential function and no variable v_k is a free variable of the function x." (p. 178)

(c) a psycho-physical product (although the second sentence points out an essential defect of this position, the use of the actions/product distinction introduced by Twardowski is remarkable):

> "Normally, expressions are regarded as the products of human activity (or as classes of such products). From this standpoint the supposition that there are infinitely many expressions appears to be obviously nonsensical." (p. 174, note 2)

(d) a physical body (we have here also critical comments):

> "But another possible interpretation of the term 'expression' presents itself: we could consider all physical bodies of a particular form and size as expressions. The kernel of the problem is then transferred to the domain of physics. The assertion of the infinity of the number of expressions is then no longer senseless and even forms a special consequence of the hypotheses which are normally adopted in physics or geometry." (p. 174, note 2)

Although Tarski had serious reservations with respect to (c) and (d) concerning the number of admissible formulas, he still was sympathetic to considering language as finitistic (observe again the importance of the distinction between act and products) (1933, p. 253, note 1):[110]

---

[109] Tarski commenting on his view about expressions as tokens added the following note in 1956 (see Tarski 1956, p. 62): "This [...] expresses the views of the author when this article was originally published and does not adequately reflect his present attitude". Although this formulation is slightly cryptic, one can assume that Tarski alludes here to his transition to the view that linguistic expressions are types.

[110] It appears that this question was very important for logicians in Warsaw in the interwar period. Łukasiewicz (1936, p. 240) notes the tension between the fact that we have only a finite number of expressions and our need in logic for infinitely many formulas. Doubtless this question must have been discussed in Warsaw, and Tarski was the first who mentioned it in print.

In the course of our investigation we have repeatedly encountered [...] the impossibility of grasping the simultaneous dependence between objects which belong to infinitely many semantical categories; the lack of terms of 'infinite order'; the impossibility of including in one process of definition, infinitely many concepts; and so on [...]. I do not believe that these phenomena can be viewed as a symptom of the formal incompleteness of the actually existing languages-their cause is to be sought rather in the nature of language itself: language, which is a product of human activity, necessarily possesses a 'finitistic' character, and cannot serve as as adequate tool for the investigation of facts, or for the construction of concepts, of an eminently 'infinitistic' character.

Nominalism

As we noted at the beginning of this chapter, nominalism is one of the most intriguing of Tarski's views. It was clearly stated in Mostowski's chapter on Tarski (see also Suppes 1988 for Tarski's caution in announcing his philosophical views) (Mostowski 1967, p. 81):[111]

> Tarski, in oral discussions, has often indicated his sympathies with nominalism. While he never accepted the "reism" of Tadeusz Kotarbinski, he was certainly attracted to it in the early phase of his work. However, the set theoretical methods that form the basis of his logical and mathematical studies compel him constantly to use the abstract and general notions that a nominalist seeks to avoid. In the absence of more extensive publications by Tarski on philosophical subjects, this conflict appears to have remained unresolved.

Some sources clearly confirm Tarski's pro-nominalist position. On April 29–30, 1965, he was chairing the joint meeting (held in Chicago) of the Association for Symbolic Logic and the American Philosophical Association on the philosophical implications of Gödel's incompleteness theorems. Tarski's remarks are preserved on tape. He said (Feferman and Feferman 2004, p. 52):

> I happen to be, you know, a much more extreme anti-Platonist. [...] However, I represent this very [c]rude, naïve kind of anti-Platonism, one thing which I would describe as materialism, or nominalism with some materialistic taint, and it is very difficult for a man to live his whole life with this philosophical attitude, especially if he is a mathematician, especially if for some reasons he has a hobby which is called set theory.

Other similar remarks by or about Tarski are collected in Feferman and Feferman (2004); these quotations are taken from Tarski's speech at the celebration of his 70th birthday as remembered by Chihara, Chateaubriand and the Fefermans themselves:

> I am a nominalist. This is a very deep conviction of mine. It is so deep, indeed, that even after my third reincarnation, I will still be a nominalist. [...] People have asked me, 'How

---

111 It is interesting that Mostowski himself was also attracted by reism, at least on special occasions, namely when he encounters very abstract constructions in set theory. See Kotarbińska (1984, p. 73). Let us add that Tarski was ready to discuss philosophical matters in conversations and seminars.

can you, a nominalist, do work in set theory and logic, which are theories about things you do not believe in?' ... I believe that there is a value even in fairy tales. (p. 52)
[I am] a tortured nominalist. (p. 52)
Elsewhere Tarski has said more specifically that he subscribed to reism or concretism (a kind of physicalistic nominalism) of his teacher Tadeusz Kotarbiński. (p. 352, note 10)

Note, however, that Tarski, contrary to Kotarbiński, never based his nominalism or reism on Leśniewski's system LO. On the other hand, we should note that the Feferman's statement contradicts Mostowski's claims. See also Mycielski (2004, pp. 215–217) about Tarski's nominalistic sympathies.

Fortunately, we can now say more about Tarski's sympathies to nominalism. This is possible due to the discovery of Carnap's protocols from the discussions between him, Tarski, Quine and (occasionally) Russell at Harvard in the early 1940s.[112] Carnap recorded the following remarks on nominalism and finitism (Mancosu 2005, p. 342):

> Tarski: At bottom, I only understand a language that fulfills the following conditions:
> 1. Finite number of individuals.
> 2. Realistic (Kotarbiński): The individuals are physical things.[113]
> 3. Non-Platonic: Only variables for individuals (things) occur, not for universals (classes etc.)

The following exchange is also recorded (Mancosu 2005, p. 334):

> I [Carnap]: Should we construct the language of science with or without types?
> He [Tarski]: Perhaps something else will emerge. One would hope and perhaps conjecture that the whole general set theory, however beautiful it is, will in the future disappear. With the higher types Platonism begins. The tendencies of Chwistek and others ('Nominalism') of speaking only of what can be named are healthy. The problem is only how to find a good implementation.

Mancosu also reports this summary, by Carnap, of views of Tarski's Polish predecessors and Tarski's own shift away from them, as he learned of them from Tarski (2005, pp. 333–334):

> The Warsaw logicians, especially Leśniewski and Kotarbiński saw a system like PM [*Principia Mathematica* – our remark, R.M., J.W.] (but with simple type theory) as the obvious system form. This restriction influenced strongly all the disciples; including Tarski until the 'Concept of Truth' (where the finiteness of the levels is implicitly assumed and neither transfinite types nor systems without types are taken into consideration; they are discussed

---

112 These protocols are in the Rudolf Carnap Collection in Pittsburgh. We are using here Frost-Arnold (2004) and Mancosu (2005) and quoting after them.
113 Frost-Arnold (2004, p. 278) adds here: "Later, Tarski relaxes this requirement: the number of individuals is allowed to be infinite or finite; neither is assumed". Mancosu (2005, p. 343) writes that this condition should be corrected to read "reistic" as opposed to "realistic". This makes sense on account of the reference to Kotarbiński.

only in the Postscript added later). Then Tarski realized that in set theory one uses with great success a different system form. So he eventually came to see this type-free system form as more natural and simpler.

Although all this indicates Tarski's decisive sympathies towards nominalism, reism and so on, we should note once again the dissonance in Tarski's views, namely between his logical and mathematical practice and some of his philosophical views; Tarski himself was aware of this situation as the quoted passages show. To understand Tarski's attitude, one should take into account the attitude of Polish mathematicians and logicians (see Murawski 2004a for a more extensive treatment of this question). According to it one should study problems using any fruitful methods and making no philosophical presuppositions. There is no need to announce one's philosophical views concerning the investigated problems because this does not belong to scientific duty, this is a "private" affair. Tarski's attitude was in full accordance with this. To some extent, he followed the pattern of doing philosophy in the Lvov–Warsaw School. Twardowski and his students distinguished "metaphysicism", i.e. limiting concrete research by metaphysical assumptions, from genuine scientific work. Although in philosophy this attitude is even more difficult to maintain, if it can be maintained at all, than in mathematics, it had an important influence on Tarski.

Language and meaning

We believe that one of the most important of Tarski's philosophical remarks about the background of the semantic theory of truth is this (1933, pp. 166–167):

> It remains perhaps to add that we are not interested here in 'formal' languages in sciences in one special sense of the word 'formal', namely sciences to the signs and expressions of which no material sense is attached. For such sciences the problem here discussed [the problem of truth – our remark, R.M., J.W.] has no relevance, it is not even meaningful. We shall always ascribe quite concrete and, for us, intelligible meanings to the signs which occur in the language we shall consider. The expressions which we call sentences still remain sentences after the signs which occur in them have been translated into colloquial language. The sentences which are distinguished as axioms seem to us materially true, and in choosing rules of inference we are always guided by the principle that when such rules are applied to true sentences the sentences obtained by their use should also be true.

We will not enter into complex issues concerning the concept of meaning, nor do we claim that Tarski defined this notion in the quoted passage or elsewhere. He did not do so, and it is well known that he avoided saying what meaning is. He believed that meanings are in language and that this is enough for a logician. The importance of Tarski's words stems from the fact that he explains what he means by formal language and that he understood that the concept of truth has no application for purely formal (syntactic) systems. Thus, the concept of interpretation is fundamental, but one must grasp meaning in order to know how signs are interpreted. According to Tarski meanings are intuitively grasped.

This view has its roots in Leśniewski. He introduced so-called intuitive formalism (we prefer this label over "intuitionistic formalism" used in the original) in the following way (1929, 487–488):

> Having no predilection for 'various mathematical games' that consist in writing out according to one or another conventional rule various more or less picturesque formulae which need not be meaningful or even – as some of the 'mathematical gamers' might prefer – which should necessarily be meaningless, I would not have taken the trouble to systematize and to often check quite scrupulously the directives of my system, had I not imputed to its theses a certain specific and completely determined sense, in virtue of which its axioms, definitions, and final directives [...] have for me an irresistible intuitive validity. I see no contradiction therefore, in saying that I advocate a rather radical 'formalism' in the construction of my system even though I am an obdurate 'intuitionist'. Having endeavoured to express my thoughts on various particular topics by representing them as a series of propositions meaningful in various deductive theories, and to derive one proposition from others in a way that would harmonize with the way I finally considered intuitively binding, I know no method more effective for acquainting the reader with my logical intuitions than the method of formalizing any deductive theory to be set forth. By no means do theories under the influence of such formalizations cease to consist of genuinely meaningful propositions which for me are intuitively valid. But I always view the method of carrying out mathematical deduction on an 'intuitionistic' basis of various logical secrets as considerably less expedient method.

However, Leśniewski's view about the role of intuitive grasping of meaning extended views of Twardowski. Let us recall that every mental act is intentional; and it has a content which is obvious to the acting mind, whatever is mental. In this respect, there was no difference between mentalism and reism. Since, to repeat once again, the meanings of linguistic expressions are objectivized mental contents, intentionally directed to objects, immediate and intuitive grasping of them (meanings) is comprehensible. Moreover, semantic properties of expressions derive from the intentional character of acts. This means that the essential features of linguistic activities are displayed adequately by properties of the corresponding expressions (see Woleński 2002 for a more detailed account). This theoretical scheme, though incomplete as a theory of meaning, functions well as an explanation of how interpretations come in. In particular, there is no conflict between formalized and interpreted languages.[114]

Note. The financial support of the Foundation of Polish Science for Roman Murawski during the writing of this paper is acknowledged.

---

114 Sinaceur (2009) introduces the term "semantic formalism" in order to capture Tarski's position. This term seems to us quite apt.

Benedykt Bornstein's Philosophy of Logic and Mathematics

Benedykt Bornstein was a significant Polish philosopher who now is almost completely forgotten. Although he wrote his doctoral dissertation under the supervision of Kazimierz Twardowski, the founder of the famous Lvov–Warsaw School of Philosophy,[115] he was not a member of this school – mainly because of his metaphysical views. In some way he was an individualist; his research did not follow the main trend.

Bornstein was born in Warsaw on 31 January 1880. He studied in Warsaw and Berlin. In 1907, he received his doctoral degree at the University of Lvov under the supervision of Kazimierz Twardowski. From 1915 he lectured on logic, epistemology and ontology within the framework of the Warsaw Society of Science Courses and from 1918 in the Free Polish University (Polish: Wolna Wszechnica Polska). From 1928 he also worked in the Łódź branch of the Free Polish University. After the Second World War, he held the Chair of Logic and Ontology at the University of Łódź. He died suddenly after a surgery in Łódź on 11 November 1948.

Bornstein's scientific interests were on the border of philosophy and mathematics. His conceptions did not win recognition and greater interest of his contemporaries. He worked in relative isolation although he participated in philosophical congresses and published his works in the major periodicals both in Poland (such as *Przegląd Filozoficzny, Wiedza i Życie, Przegląd Klasyczny*) and abroad. His scientific activities can be divided into three periods: in the first one he translated Kant's works and developed his ideas in a critical way; the second period was dedicated to investigations concerning the philosophy of mathematics, and the third period – to problems of metaphysics cultivated in the spirit of the classical trend. His works written in the second period raised some interest of Polish philosophers. His investigations concerning the philosophy of mathematics led to the formulation of a new philosophical method in the form of categorial geometrical logic. The theme of this paper makes us focus on the latter investigations.

Let us begin by discussing Bornstein's reflections on the philosophy of geometry. Here Bornstein referred to Kant's transcendental aesthetics and Twardowski's theory of images and concepts (cf. his 1894). At the same time, he criticized the idea of constructing geometry on the basis of set theory or topology; he also distanced himself from Poincaré's conventionalism. In his opinion, constructing a geometry

---

115 For the Lvov–Warsaw School of Philosophy see the monograph by Woleński (1989).

should be begun by constructing proper geometrical concepts, which have their objective references. In his book *Prolegomena filozoficzne do geometryi* [Philosophical Prolegomena to Geometry] (1912) he distinguished between the image of physical space and the concept of geometrical space, and he followed the idea that the so-called background image must be an image the object of which exists and is truly perceived, which is to guarantee that the common features of the object of the concept of geometrical space and the object of the background image will not only concern the world of objective images but also be grounded in the experiential reality.[116] According to Bornstein, one of the common features of both objects is three-dimensionality. He wrote in *Prolegomena filozoficzne do geometryi*:

> If we analyse this image with respect to spatiality we will be always convinced that its object is three-dimensional, i.e. it has length, width and height (or depth); that from each of its points we can draw three perpendicular lines, belonging to the given object in some space. This objective spatiality, characterising three-dimensionality, is a common feature of our background image and the object of the concept of geometrical space, based on that image (1912, p. 8).[117]

---

116 It is worth mentioning here Leśniewski's ideas concerning the distinction between objects and concepts. The reason is that both Leśniewski and Bornstein were students of Twardowski, so they came from the same philosophical school. Below we shall present the discussion between Leśniewski and Bornstein dealing with the foundations of set theory. It is the other reason to consider Leśniewski's ideas. Leśniewski spoke not about concepts but about names (in particular about general names) and about individual objects that could be of arbitrary nature. Further one should distinguish object in his ontology (it was in fact a calculus of names) and in his mereology (that was the theory of sets in the collective sense). According to Leśniewski a name is any expression which can play a role of B in sentences of the form "a is B". Hence Leśniewski proposed in his ontology a theory of names of one category only and liquidated the dualism of nominal expressions (individual names vs. general names). He did not say anything on the nature of objects that exist except that they are individual objects. His ontology is "metaphysically" neutral – it cannot be deduced from its theses whether anything does exist and what does exist. He says only that A exists if and only if for some x, x is A, and that A is an object if and only if for some x, A is x. Add that on Leśniewski's ontology were founded ontological considerations of Tadeusz Kotarbiński. His conception is called reism – it is claimed in it that individual objects are things and that only they do exist.

117 „Jeżeli zanalizujemy takie wyobrażenie pod względem przestrzenności przekonamy się zawsze, że przedmiot jego jest trójwymiarowy, t.j. że posiada długość, szerokość i wysokość (względnie głębokość), że w każdym jego punkcie można poprowadzić trzy prostopadłe linie, należące na pewnej przestrzeni do danego przedmiotu. Ta przestrzenność przedmiotowa, którą charakteryzuje trójwymiarowość, jest cechą wspólną przedmiotu naszego wyobrażenia podkładowego i przedmiotu pojęcia przestrzeni geometrycznej, opartego na tem wyobrażeniu".

Three-dimensionality is determined by experience and is not – as Poincaré claimed – a separate mental construction.[118]

As far as the question of the choice between Euclidean and non-Euclidean geometries is concerned, Bornstein thought that:

> From the purely logical or analytical point of view the theorems or formulas of non-Euclidean geometry contain no contradictions, and it is logically possible that they are equally eligible as the theorems and formulas of Euclidean geometry (1912, p. 89).[119]

At the same time, experience cannot help us to choose one, true and correct geometry. Bornstein wrote in *Prolegomena*:

> If now the followers of the purely logical or analytical concept of geometry turn to experience with the question which of the three logically possible systems of theorems is important to experience and is confirmed by it, they must be prepared not to receive any answer to their question. […] In other words, when we turn to experience to show us which of the possible logical systems is confirmed by it, which is true, then experience will never give us any answer since its data will present a constant in the equation with two unknowns (one geometrical and the other physical), and so they will be insufficient to solve precisely this geometrical unknown in the equation (1912, pp. 89–90).[120]

Bornstein claimed that real spatial extensiveness could not be identified with the extensiveness defined by the continuum of real numbers. The latter has no space character. Therefore, the attempts to transfer theorems from one domain to the other are not justified. In particular, one cannot assume *a priori* that a geometrical line does not correspond to any continuous function. In his article „Problemat istnienia linji geometrycznych" [The Problem of the Existence of Geometrical Lines] (1913) he showed that such lines corresponded to some solid functions and did

---

118 For the particular remarks on Bernstein's views concerning the problem of essence and structure of geometrical space, see Śleziński (2009).
119 „Z punktu widzenia czysto logicznego lub czysto analitycznego twierdzenia lub formuły geometryi nieeuklidesowej nie zawierają sprzeczności, a logicznie możliwe, są równie uprawnione, jak twierdzenia i formuły geometryi euklidesowej".
120 „Jeżeli teraz zwolennicy czysto logicznego lub czysto analitycznego pojmowania geometryi zwrócą się do doświadczenia z pytaniem, który z trzech logicznie możliwych systemów twierdzeń jest ważny dla doświadczenia i znajduje w nim potwierdzenie, to muszą być przygotowani na to, że odpowiedzi na to pytanie nie otrzymają. […] Słowem, gdy zwracamy się do doświadczenia, by nam wskazało, który z możliwych logicznie systemów znajduje w nim potwierdzenie, który jest prawdziwy, to doświadczenie na to pytanie nigdy nie będzie mogło dać nam odpowiedzi, gdyż jego dane będą przedstawiały wielkość stałą w równaniu z dwiema niewiadomymi (jedną geometryczną, drugą fizyczną), a więc będą niedostateczne do ścisłego rozwiązania tego równania co do niewiadomej geometrycznej".

not correspond to other ones. Assuming that all geometrical curves have tangents, we have the result that only functions with derivatives correspond to them. Consequently, if every movement must have speed, and speed is the derivative of distance with respect to time, movement cannot occur along curves without tangents. Thus not all functions are of geometrical character, in particular it concerns those functions that have no derivatives.

Bornstein also dealt with the problem of infinity. In his opinion, an infinite set can be given only as a certain whole embracing infinitely many elements. At the same time, the actual infinity is never given as the infinity of its particular elements – only a finite number of them can actually be given.[121] Thus, a question arises whether all elements of an infinite set (in the sense of actual infinity) exist physically or whether they exist in themselves independently from their actualization. Bornstein examined these questions in his book *Elementy filozofii jako nauki ścisłej* [Elements of Philosophy as an Exact Science] (1916) asking whether an actual segment is a set of potential or actual points. He concluded that an infinite set of points situated between two points of a geometrical line existed physically in nature but not all of its elements necessarily did.

Thus we come to Bornstein's considerations on the foundations of set theory. We must above all mention his work „Podstawy filozoficzne teorji mnogości" [The Philosophical Foundations of Set Theory] (1914). This work was criticized by Stanisław Leśniewski in his article „Teorja mnogości na "podstawach filozoficznych" Benedykta Bornsteina" [Set Theory on the 'Philosophical Foundations' of Benedykt Bornstein] (1914). In turn, Bornstein wrote an article „W sprawie recenzji p. Stanisława Leśniewskiego rozprawy mojej pt. 'Podstawy filozoficzne teorji mnogości'" [On Mr Stanisław Leśniewski's Review of My Dissertation 'The Philosophical Foundations of Set Theory'] (1915). Thus the polemic ended.

We cannot discuss here the technical details of the polemic and more, the polemic did not bring about any effects. However, some arguments of both thinkers are worth mentioning.

Let us begin by stating that in his work (1914) Bornstein notices that the source of antinomy in set theory is its erroneous philosophical justification. He concludes that a set of individually existing elements can be only finite. In addition, he bases his thesis concerning the existence of finite and infinite sets having individually existing elements on the following three lemmata (cf. 1914, pp. 183–185):

- The same number corresponds to two equivalent sets with individually existing elements.

---

121 Observe that Bornstein's idea of an infinite set is not identical with Cantor's one. Cantor – following Platonizm – did not distinguish between existing and actually given elements.

- In a set of elements, existing individually, the same number cannot correspond to the proper part of this set in the same way as to the whole.
- A set of elements, existing individually, cannot be equivalent to its own part.

He explains the used terms in the following way:

> If a plurality of elements, each existing individually, i.e. as a different unit, is analysed only as a plurality of units, we analyse it from the point of view of quantity; at the same time, this plurality of units constitutes the quantity, relatively, its number of individually existing elements of the given plurality. [...] between the plurality of elements, existing individually, and the plurality of units, constituting its quantity, relatively its number, there is one–one correspondence; these pluralities are, as we say, equivalent or of equal power. [...] since quantity is a real feature of the plurality of elements, existing individually, whereas the number is a notional equivalent of this feature (1914, p. 183).[122]

Omitting the technical details of Bornstein's reasoning, we must say that he made the error of *quaternio terminorum*, i.e. the use of the same term in two different meanings – in this case it is the term "the same number".

Assuming the existence of an infinite set of natural numbers, Bornstein shows the essential nature of infinite pluralities. Now, in the infinite plurality of natural numbers only their finite quantity – in his opinion – can be considered individually. Therefore, there can be infinite pluralities without any possible individual content. He writes:

> [...] here we have a perfect example, showing the essential nature of infinite pluralities, consisting in their full independence from the matters of actualising (individualising, materialising) the elements of plurality. Here we have an example of a pure form in ideal perfectness (1914, p. 190).[123]

He also concludes that the well-ordering theorem (equivalent to the AC) "applying in general to all kinds of plurality is wrong; whereas applying to the plurality

---

[122] „Jeżeli mnogość elementów, z których każdy istnieje indywidualnie, tj. jako różna od innych jednostka, rozpatrujemy tylko jako mnogość jednostek, to rozpatrujemy ją z punktu widzenia ilości, przy czym ta mnogość jednostek stanowi właśnie ilość, względnie liczbę istniejących indywidualnie elementów danej mnogości. [...] między mnogością elementów, istniejących indywidualnie, a mnogością jednostek, stanowiącą jej ilość, względnie liczbę, istnieje odpowiedniość jedno-jednoznaczna; mnogości te są, jak mówimy, równoważne lub równej mocy. [...] ilość bowiem jest cechą rzeczywistą mnogości elementów istniejących indywidualnie, liczba zaś jest odpowiednikiem pojęciowym tej cechy".

[123] „[...] mamy tu doskonały przykład, wykazujący istotną naturę mnogości nieskończonych, polegającą na ich zupełnej niezależności od spraw zaktualizowania (zindywidualizowania, zmaterializowania) elementów mnogości. Mamy tu przykład czystej formy w idealnej doskonałości".

of elements, existing individually, physically, is an obvious truth" (Bornstein 1914, p. 190).[124]

Leśniewski began his criticism of Bornstein's work (1914) with the following words:

> Dr Benedykt Bornstein wrote a treatise in which he tried to provide set theory with 'philosophical foundations'; he thought that certain contradictions, which can be seen in set theory, are not caused by set theory but by its wrong philosophical justification, and this view of the problems, prevailing in set theory, must have been the origin of the author's desire to add to this science some thoughts, which could justify it 'philosophically' (Leśniewski 1914, p. 488).[125]

Further, Leśniewski analyses Bornstein's formal argumentations – ignoring the ontological questions, which were so important to the latter. In particular, Leśniewski criticizes Bornstein's terms "existing individually" and "existing formally", accusing him of not giving any precise definition of the concept of "unit". In addition, he proposes to replace the term "unit" by the term "object", which, however, as seen in Bornstein's response (1915) does not satisfy the latter. Leśniewski also criticizes Bornstein's interpretation of Zermelo's well-ordering theorem.

Avoiding any complicated (and devoid of deeper meaning now) technical questions concerning the polemic between Leśniewski and Bornstein, it would be sufficient to say that their levels of discourse were entirely different. Leśniewski defended the standard approach towards set theory (which he then refuted for the cause of mereology) against Bornstein's criticism flowing from philosophical motives. As Śleziński (2010) notices: "for Leśniewski the formal analyses are binding whereas for Bornstein the argumentations, apart from formal correctness, must refer to the objective layer of the problems under consideration" (p. 110).[126] Leśniewski summarized his critical review of Bornstein's words in the following way:

> The work of Mr Bornstein has no value for the 'foundations' of set theory. It does not remove any 'contradictions' from set theory as Mr Bornstein seems to be claiming; on the

---

124 „w zastosowaniu do wszelkiej mnogości w ogóle jest błędne; w zastosowaniu natomiast do mnogości elementów, istniejących indywidualnie, aktualnie, jest prawdą oczywistą".

125 „Dr Benedykt Bornstein napisał rozprawę, w której starał się zaopatrzyć teorię mnogości w „podstawy filozoficzne"; uważał on, iż do pewnych sprzeczności, które dają się widzieć w teorii mnogości, prowadzi nie sama teoria mnogości, lecz błędne jej uzasadnienie filozoficzne, a pogląd taki na stan rzeczy, panujący w teorii mnogości, stanowił właśnie zapewne genezę pragnienia autora, by przysporzyć tej nauce trochę myśli, które by ją mogły „filozoficznie" uzasadnić".

126 „dla Leśniewskiego wiążące są analizy formalne, a dla Bornsteina rozumowania, oprócz poprawności formalnej, muszą odnosić się do warstwy przedmiotowej badanych problemów naukowych".

contrary, he creates them to a much bigger extent; he does not justify them 'philosophically' and in no other way does he justify even one theorem of set theory; since one cannot justify something with the help of 'definitions' and 'lemmata' that are full of errors and contradictions; he explains nothing because the seemingly devised conceptions of something, for example the conception of 'capacity,' are inconsistent and unclear (1914, p. 507).[127]

In his response (Bornstein 1915) to Leśniewski's criticism, Bornstein tried to specify his conception of set theory. He also saw certain inconsistencies in Leśniewski's arguments. He was not convinced about the validity of the accusations and concluded his answer:

> Facing the foregoing arguments it seems to me that I will be objective responding to Mr Leśniewski's review: *primo* – it does not show, even to the slightest extent, any contradictions which are to be stuck in the concepts I have used, and *secundo* – it is an example of Mr Leśniewski's extremely careless disregard of the elementary principles of logic (Bornstein 1915, pp. 139–140).[128]

As we have seen, both debaters remained on different planes. Leśniewski conducted his argumentation and analyses in the spirit preferred by the Lvov–Warsaw School, i.e. using the apparatus of mathematical logic and focusing on formal matters, whereas Bornstein favoured ontological questions and worked in the spirit of the concept of the mathematics of quality, which he was developing himself. In particular, the latter might have been the reason why there were no polemics (except the one held by Leśniewski) with Bornstein's later works – in fact, the concept of the mathematics of quality was so different from the universally accepted tendencies and styles of thinking that it was difficult to find any common points. On the other hand, Bornstein criticized the widespread practice of treating mathematics as the science about quantity and magnitude, number and measure – in his opinion there is also qualitative mathematics, especially qualitative algebra or geometry. This qualitative mathematics deals not only with order, in particular with order between qualities. It should serve a mathematization of the philosophy and

---

[127] „Praca p. Bornsteina nie ma żadnej w ogóle wartości dla „podstaw" teorii mnogości. Nie usuwa ona żadnych „sprzeczności" z teorii mnogości, jak się to zdaje p. Bornsteinowi, lecz je przeciwnie w wielkiej obfitości stwarza; nie uzasadnia „filozoficznie" ani też w żaden inny sposób ani jednego twierdzenia teorii mnogości, nie można bowiem uzasadnić czegoś za pomocą „definicji" i „lematów", pełnych błędów i sprzeczności; nie wyjaśnia nic, bo obmyślone niby czegoś koncepcje, jak np. koncepcje „pojemności", są sprzeczne i niejasne".

[128] „Wobec powyższego wydaje mi się, że będę obiektywnym, gdy o recenzji w mowie będącej p. Leśniewskiego powiem: *primo* – że w najmniejszym nawet stopniu nie wykazuje sprzeczności, tkwić mających w używanych przeze mnie pojęciach, i *secundo* – że jest przykładem niebywale lekkomyślnego nieliczenia się p. Leśniewskiego z elementarnymi zasadami logiki".

the construction of a qualitative-mathematical model of the world. Let us add that details of Bornstein's attempts to develop the qualitative mathematics are not quite clear.

Let us proceed to the next idea of Bornstein, namely, his conception of the geometrization of logic, i.e. geometrical logic. Referring to Leibniz, who was always closer to the intensional than the extensional conception of logical forms and who wanted to construct logic based on the content of expressions and not only on the extensions of concepts, Bornstein tried to create a new logic – namely the logic of content, since he thought that the content of a concept sets out its extension, and thus the exactness and definiteness of the content determine the precision and definiteness of the extensions and in general, of the classes.

Bornstein divided concepts and judgements into those which were set out objectively and those which were set out logically. The former are parallel to objects in reality and the latter gain their meaning through definitions. In addition, Bornstein distinguishes between nominal and real definitions. In nominal definitions the definiendum as if synthesises the essence of words constituting the definiens. In real definitions we have the reverse process – the definiendum is divided into a combination of simpler constituents occurring in the definiens. However, both types of definition are definitions *per genus proximum et differentiam specificam*. Likewise, we have judgements set out objectively and judgements set out logically. At the same time, Bornstein assumes that all judgements have subject-predicative structures.

Bornstein, following the conceptions of Edward Vermilye Huntington (1904), proposed his own system of the algebra of logic, which he formulated as categorial. He accepted three logical operators: negation, addition and multiplication. Addition consists in integrating the contents of concepts whereas multiplication sets out the biggest common element of concepts. Here two constants appear: 0 and 1, where 0 is the lower bound of any content and 1 is the upper bound of any content. Element 0 has the weakest logical content since when added to any element it does not change the content of it. Element 0 expresses the content of the concept of "something" or "the object in general" whereas element 1 presents the substantially strongest concept, concept with the richest content, "whole" and "everythingness". The element 1 is the upper limit of all concepts whereas 0 is the lower limit of all concepts. Moreover, there is a relation of the subordinance of content marked as but it does not have the property of connectedness.

Furthermore, Bornstein tried to give a geometrical interpretation to his categorial logic of content.[129] His first attempts can be found in „Zarys architektoniki i

---

[129] Add that the adjective "categorial" means here something else than in Ajdukiewicz' "categorial grammar".

geometrji świata logicznego" [Outline of Architectonics and Geometry of the Logical World] (1922), and then in his more mature work „Geometrja logiki kategorialnej i jej znaczenie dla filozofii" [Geometry of Categorial Logic and Its Importance for Philosophy] (1926).[130] However, we cannot get entangled in complicated (and not always clear) technical details. Suffice it to say that Bornstein refers to projective geometry stressing its qualitative character. He shows the structure of his logic of content through various diagrams, both two-dimensional and three-dimensional. Thus he refers to the works of the previous authors who used a geometrical exposition of certain logical dependencies, e.g. Euler's wheels, the diagrams of Venn and Hasse or certain conceptions of Leibniz, Peirce and Grassmann.

The analyses on logic and the use of geometrical interpretations led Bornstein to the conclusion that both domains could be linked and thus a qualitative-categorial geometrical logic could be created. This logic can help us discover and reveal the universal structures of reality. In his work *La logique géométrique et sa portée philosophique* [Geometrical Logic and Its Meaning for Philosophy] (Bornstein 1928), he tried to show the similarity of the domain of thought and the domain of space objects. He tried to unite both of his systems: algebraic logic and geometrical logic in one system called topologic (Polish: topologika).

Bornstein's system of qualitative-categorial geometrical logic is not quite clear – therefore we cannot go into details. Let us say only that he used in his system some ideas of projective geometry. He considered two-dimensional and three-dimensional categorial logic. The two-dimensional logico-geometrical space was spread by him on two categories: *genus proximus* and *differentia specifica*, whereas the three-dimensional one on those two categories and additionally on the category of individualization (individual determination).

Bornstein generalized his system of logic as a dialectical geometrical logic and presented it in his unpublished work *Zarys teorii logiki dialektycznej* [Outline of the Theory of Dialectical Logic] (1946). Unfortunately, his explanations were not clear enough. It should be stressed that he assumed the possibility of various degrees of dialecticality and consequently, various kinds of dialectical logics. In his opinion traditional logic is the least dialectical one whereas mathematical logic is partially dialectical. In the quoted work he wanted to show that dialectical logic could be treated in a mathematical way, could be axiomatized and given a geometrical interpretation. However, the problem of the consistency of dialectical logic appeared. The need to show consistency was very essential and more, this logic was to help examine the real world. The sought-after proof of consistency would refute the accusation of the irrationality of this logic. Unfortunately, Bornstein did

---

130 Cf. also his unpublished works (a), (b) and (c).

not give such a proof – he gave only certain arguments supporting consistency but they were disputable.

Bornstein's considerations were based on his conviction that there existed a harmony between the world of non-spatial thoughts and the world of spatial beings. He thought that mathematics and the logic of quality were objectively grounded in the real world. At the same time, he treated mathematics as an auxiliary domain of philosophy. Bornstein wanted to construct a philosophical system using mathematical concepts. He thought after certain universal structures and principles of the real world; besides the quantitative aspect he looked for the qualitative aspect. In his opinion, the order of the world concerns both of these aspects. Thus he spoke about the mathematics of quantity and the mathematics of quality. Mathematics is not only the science of quantity and measure but of order, in particular the order between qualities. For Bornstein, metrical geometry was an example of the mathematics of quantity whereas projective geometry – the mathematics of quality. Philosophy should look for the qualitative structures of the world – its starting point should be qualitative mathematical logic.

Bornstein's conceptions did not win recognition and acceptance of his contemporaries. The reasons for this included the lack of clarity and precision of his ideas. Moreover, they were not completely worked out. Bornstein's investigations did not follow the main trend of research. The mathematical and logical tools he constructed were to create a metaphysical system and not to serve analyses, which was decidedly different from the style of philosophy accepted and developed in the Lvov–Warsaw School.

Note. The financial support of the National Center for Science [Narodowe Centrum Nauki] (grant no N N101 136940) is acknowledged. The paper is based on my book *Philosophy of Logic and Mathematics in Poland in the 1920s and 1930s* (Murawski 2014) published by Birkhäuser Verlag, Basel. I would like to thank Professor Jan Woleński for helpful hints concerning Leśniewski's philosophy.

Philosophy of Logic and Mathematics in the Warsaw School of Mathematical Logic

The Warsaw School of Mathematical Logic was a part of the Lvov–Warsaw School of Philosophy. It belonged to the most important centres of mathematical logic between the wars. It is natural to ask what were the philosophical views and attitudes of logicians in Warsaw towards mathematics and logic itself. One can also ask whether and to what extent those views influenced formal and technical research, whether that research had its source in philosophical considerations or was it independent of any philosophical presuppositions. Did the philosophical views bound the technical investigations or were they without meaning for them?

The attitude of Polish logicians and mathematicians towards the philosophy of mathematics can be shortly characterized as follows: they saw the mathematical and philosophical foundations of mathematics as independent although connected in a way and indispensable for understanding logical and mathematical activity. With two exceptions (Chwistek and Leśniewski) they represented a view guided by the following two principles:

- All commonly accepted mathematical methods should be applied in metamathematical investigations.
- Metamathematical research cannot be limited by any *a priori* accepted philosophical standpoint.

On the other hand, logic and mathematics have their own genuine philosophical problems which should not be neglected. In particular, although metamathematical results do not solve philosophical controversies about mathematics and logic, yet the former illuminate the latter.

What were the sources of such an attitude? One can indicate two of them. The first one can be exemplified by Sierpiński's work on the axiom of choice (AC) and its applications in mathematics. In his paper (1918) on the role of AC, Sierpiński distinguished two independent questions:

- philosophical controversies around this axiom and
- its place in proving mathematical theorems.

According to Sierpiński, the second issue should be investigated independently of philosophical inclinations concerning the problem whether the AC is to be accepted or not. This opinion was included in all editions of Sierpiński's textbook on set theory from 1923 (*Zarys teorii mnogości* [An Outline of Set Theory], 1923) to 1965 (*Cardinal and Ordinal Numbers*, 1965). In 1965 (p. 95) he wrote:

Still, apart from our personal inclination to accept the axiom of choice, we must take into consideration, in any case, its role in the set theory and in the calculus. On the other hand, since the axiom of choice has been questioned by some mathematicians, it is important to know which theorems are proved with its aid and to realize the exact point at which the proof has been based on the axiom of choice; for it has frequently happened that various authors have made use of the axiom of choice in their proofs without being aware of it. And after all, even no one questioned the axiom of choice, it would not be without interest to investigate which proofs are based on it and which theorems are proved without its aid – this, as we know, is also done with regard to other axioms.

This means simply that one should disregard philosophical controversies (and treat them as a "private" matter) and investigate (controversial) axioms as purely mathematical constructions using any fruitful methods.

The second source of the discussed attitude of Polish mathematicians and logicians towards philosophy was the tradition of Polish analytic philosophy originated by Kazimierz Twardowski in Lvov. According to Twardowski and his students, we must clearly and sharply distinguish world-views and scientific philosophical work. This idea was particularly stressed by Łukasiewicz, the main architect of the Warsaw School of Logic. He regarded various philosophical problems pertaining to the formal sciences as belonging to world-views of mathematicians and logicians, but the work consisting in constructing logical and mathematical systems together with metalogical and metamathematical investigations constituted for him the subject of logic and mathematics as special sciences. Hence, philosophical views cannot be a stance for measuring the correctness of formal results. Yet philosophy may serve as a source of logical constructions.

One of the consequences of the described attitude of Polish logicians and mathematicians was the fact that they did not attempt to develop a comprehensive philosophy of mathematics and logic (Stanisław Leśniewski and Leon Chwistek were here the exceptions!). They formulated their philosophical opinions concerning mathematics or logic only occasionally and only on problems which just interested them or on which they actually worked. Consequently, there were in Poland no genuine philosophers of mathematics. Philosophical remarks were formulated by logicians and mathematicians only on the margin of their proper mathematical or logical works (and had no meaning for the results themselves).

The current trends and views in the philosophy of mathematics, i.e. logicism, intuitionism and formalism, were of course well known (and there appeared papers discussing those tendencies, their meaning and development). But none of them was represented in the Warsaw School. Moreover, it did not represent any other trend; it had no official philosophy of logic and mathematics. This followed from the belief that logic and mathematics are autonomous with respect to philosophy. Opinions in the field of the philosophy of logic and mathematics were treated as "private" problems and philosophical declarations were made reluctantly and seldom. If they were made then it was stressed, directly or indirectly, that these were personal opinions.

Though some of the logical investigations were motivated by philosophical problems – e.g. the many-valued logics by Łukasiewicz – the formal, logical constructions were always separated from their philosophical interpretations. Another example is the investigation on intuitionistic logic carried out among others by Tarski without accepting intuitionism as the philosophy of mathematics. The program of Janiszewski (1917) and the Polish School of Mathematics created set-theoretical foundations of mathematics in a methodological and not philosophical sense.

What were the separate philosophical opinions formulated by Polish logicians, philosophers and mathematicians? We shall answer this question considering the philosophical views of two representatives of the Warsaw School of Mathematical Logic: Alfred Tarski (1901–1983) and Andrzej Mostowski (1913–1975). Tarski belonged to the first generation of the Warsaw School and Mostowski, to the second generation.[131]

Alfred Tarski was interested in philosophical problems and very actively participated in the philosophical life of his time. He was convinced of the philosophical significance of his works, in particular of his work on truth (1933). He described himself as (cf. Tarski 1944):

> Being a mathematician (as well as a logician, perhaps a philosopher of a sort) [...]

Tarski's philosophical attitude was anti-metaphysical; he supported the idea of scientific philosophy. He accepted a program of "small philosophy" which aims at detailed and systematic analysis of the concepts used in philosophy. Such a philosophy is minimalistic, anti-speculative and sceptical towards many fundamental problems of traditional philosophy. This attitude was inherited by Tarski from the Lvov–Warsaw School and strengthened by contacts with the Vienna Circle. He maintained also empiricism and abandoned the analytic/synthetic distinction. He stressed that logical and empirical truths belong to the same generic category. Influenced by Leśniewski and Kotarbiński he was inclined to a rather strongly nominalistic understanding of expressions. One finds many places in which he confirmed this. For example, during a symposium organized by the Association for Symbolic Logic and the American Philosophical Association held in Chicago

---

[131] For logic and philosophy of logic and mathematics in Poland between the wars, see the basic monograph of Woleński (1989) as well as Woleński (1992, 1993, 1995), Murawski and Woleński (2008), Murawski (2004a, 2011a, 2014).

on 29th–30th April 1965 and devoted to the philosophical implications of Gödel's incompleteness theorems, he said (cf. Feferman and Feferman 2004, p. 52):

> I happen to be, you know, a much more extreme anti-Platonist. [...] However, I represent this very [c]rude, naïve kind of anti-Platonism, one thing which I would describe as materialism, or nominalism with some materialistic taint, and it is very difficult for a man to live his whole life with this philosophical attitude, especially if he is a mathematician, especially if for some reasons he has a hobby which is called set theory.

In the biography of Tarski written by Fefermans (2004) one finds more such quotations, e.g. (p. 52):

> I am a nominalist. This is a very deep conviction of mine. It is so deep, indeed, that even after my third reincarnation, I will still be a nominalist. [...] People have asked me: 'How can you, a nominalist, do work in set theory and logic, which are theories about things you do not believe in?' ... I believe that there is a value even in fairy tales.
> [I am] a tortured nominalist.

Elsewhere Tarski has said more specifically that he subscribed to reism or concretism (a kind of physicalistic nominalism) of his teacher Tadeusz Kotarbiński. Mostowski wrote about Tarski so (cf. 1967, p. 81):

> Tarski, in oral discussions, has often indicated his sympathies with nominalism. While he never accepted the 'reism' of Tadeusz Kotarbiński, he was certainly attracted to it in the early phase of his work. However, the set-theoretical methods that form the basis of his logical and mathematical studies compel him constantly to use the abstract and general notions that a nominalist seeks to avoid. In the absence of more extensive publications by Tarski on philosophical subjects, this conflict appears to have remained unresolved.

Tarski was inclined to identify mathematics with the deductive method. He maintained that there is no hard borderline between formal and empirical sciences. He admitted the rejection of logical and mathematical theories on empirical grounds. He claimed also that there is no sharp demarcation between logical and factual truth and that the concept of tautology is unclear.

One must stress that all those were his "private" philosophical views which did not influence his logical and mathematical research; in other words, his research was independent of any philosophical presuppositions. In the paper "Über einige fundamentale Begriffe der Methodologie der deduktiven Wissenschaften" (1930) he explicitly wrote:

> [...] it should be noted that no particular philosophical standpoint regarding the foundations of mathematics is presupposed in the present work.

This was typical for him and for the whole Warsaw School of Logic. This independence of logical and mathematical studies and philosophical views explains the cognitive conflict and discrepancy between Tarski's nominalistic and empiricistic sympathies and his "Platonic" mathematical and logical practice. Note that his attitude enabled him to contribute to various important foundational streams without the necessity of accepting their philosophical assumptions and attempting to reconcile the philosophy and the research practice. His program of metamathematics can be summarized by his words from the paper (1954) where he wrote:

> As an essential contribution of the Polish school to the development of meta- mathematics one can regard the fact that from the very beginning it admitted into metamathematical research all fruitful methods, whether finitary or not.

Andrzej Mostowski inherited his general philosophical attitude from Tarski. He freely used infinitary methods and strongly insisted that no formal work should be limited by philosophical assumptions. However, it seems that Mostowski felt himself obliged to a more extensive and systematic treatment of his views in the philosophy of mathematics. In the review of Mostowski's *Thirty Years of Foundational Studies* (1965) published in *Studia Logica*, R. Suszko characterized him as a "mathematician-logician, to whom the philosophical aspect of logic and the theory of the foundations of mathematics is not alien" (Suszko 1968, p. 169). In many of his technical papers and works, Mostowski stressed in the introductory sections or prefaces the importance and indispensability of certain philosophical presuppositions. He discussed also the possible philosophical consequences of technical mathematical results presented there. But such comments and remarks were always reduced to a minimum and had no influence on the technical considerations.

In the Introduction to the monograph *Teoria mnogości* [Set Theory] written together with K. Kuratowski[132] they wrote (1952, p. vi):

> There exists so far no comprehensive philosophical discussion of basic assumptions of set theory. The problem whether and to what extent abstract concepts of set theory (and in particular of those parts of it in which sets of very high cardinality are considered) are connected with the basic notions of mathematics being directly connected with the practice has not been clarified so far. Such an analysis is needed because by Cantor, the inventor of set theory, basic notions of this theory were enwraped with a certain mysticism.

On the other hand, the authors are convinced that the meaning and importance of set theory for the foundations of mathematics were demonstrated also in connection with the philosophy of mathematics.

And they declare that the most important feature of set theory is the fact that it provides a tool for other parts of mathematics which are directly connected with applications.

---

132 The discussed remarks were reprinted in the second and third Polish editions of the book.

The philosophical remarks were made only in Introduction. One finds in the book no further philosophical declarations or statements. In the whole book, the authors strictly distinguish (in the spirit of Sierpiński) the philosophy of the AC and its role in mathematics and set theory itself – all theorems in which AC is used are marked by a small circle.

Mostowski considered also philosophical problems in connection with Gödel's incompleteness theorems. As in the case of set theory, he indicated only the philosophical problems connected with the discussed mathematical issues and showed the possible solutions but avoided any fixed and definite philosophical declarations. Moreover, the philosophical comments were reduced to a minimum.

He stressed that we do not have a precise notion of a correct mathematical proof. In the paper (1972, p. 83) Mostowski emphasizes that: "A mathematical proof is something much more complicated than a simple succession of elementary rules contained in the so-called inference rules. [...] Therefore one must necessarily show moderation in stressing the role of logical rules in [mathematical] proofs". On the other hand, the author is sure that despite the fact that the old program of formalization of mathematics has been practically waived, "the collaboration of logic and mathematics was fruitful and probably will still bring important results" (p. 83).

Note also that the three trends in the philosophy of mathematics which dominated in the 1920s and 1930s of the 20th century (logicism, intuitionism and formalism) were the starting point of Mostowski's series of lectures *Thirty Years of Foundational Studies* (1965). He stressed there that they gave rise to the development of three directions in logico-mathematical investigations: constructivism, metamathematical and set-theoretical ones. But in the main text one finds no further philosophical remarks.

So far we have shown that Mostowski was aware of philosophical problems connected with mathematics but avoided making any explicit philosophical declarations. There is, however, one paper by him in which he makes explicit declarations, namely the paper *The present state of investigations of the foundations of mathematics* (1955a and 1955b). Unfortunately there is a problem of interpretation: the paper was written in the first half of the 1950s and the ideological atmosphere of that time could have had an influence on it. It is not possible now to decide to what extent outside factors influenced the paper. On the other hand, the author could have restricted himself to purely mathematical issues and avoided entirely any philosophical remarks and declaration. If he did not do it we can treat his remarks as genuine.

He states there (cf. 1955a, p. 42):

> [...] An explanation of the nature of mathematics does not belong to mathematics, but to philosophy, and it is possible only within the limits of a broadly conceived philosophical

view treating mathematics not as detached from other sciences but taking into account its being rooted in natural sciences, its applications, its associations with other sciences and, finally, its history.

Investigations on the foundations of mathematics by mathematical methods affect the formation of a broader philosophical view. The results obtained there confirm – according to Mostowski (cf. 1955a, p. 42):

> the assertion of materialistic philosophy that mathematics is in the last resort a natural science, that its notions and methods are rooted in experience and that attempts at establishing the foundations of mathematics without taking into account its originating in natural sciences are bound to fail.

Hence Mostowski represents here an empirical point of view in the philosophy of mathematics. As mentioned above, it is not quite clear what was the influence of outside factors (in particular of the then dominant ideology) on those views. Specific expressions used by him may suggest such an influence. Such statements could be, at least partially, a price that should have been payed to the official philosophy. On the other hand, note that empirical (or quasi-empirical) trends are since the 1960s of the last century still more and more vivid in the philosophy of mathematics.

Mostowski admitted in various places that constructivism (especially its aims, not necessarily its solutions) was always very attractive to him (cf. 1959, p. 192). The reason for that was the fact that (cf. Mostowski 1959, p. 192):

> it wants to inquire into the nature of mathematical entities and to find a justification for the general laws which govern them, whereas platonism takes these laws as granted without any further discussion.

He stressed that constructivistic trends in the foundations of mathematics are nearer to nominalistic philosophy than to the idealistic (in the Platonic sense) one. This nominalistic character implies that constructivism does not accept the general notions of mathematics as given but tries to construct them. "This leads to the result that one can identify mathematical concepts with their definitions" (Mostowski 1959, p. 178). The advantage of nominalism is the fact that several important mathematical theories have been reconstructed in a satisfactory way on a nominalistic basis, and those reconstructions have turned out to be equivalent to the classical theories.

* * *

Polish logicians and mathematicians being convinced of the importance of philosophical problems and knowing quite well the current philosophical trends treated logic and mathematics as autonomous disciplines independent of philosophical reflection on them, independent of any philosophical presuppositions.

Therefore they sharply separated mathematical and logical research practice and philosophical discussions concerning logic and mathematics. Philosophical views and opinions were treated as a "private" matter that should not influence the mathematical and metamathematical investigations. On the contrary, in the latter all correct methods could and should be used. This "methodological Platonism" enabled Polish logicians and mathematicians to work in various areas without being preoccupied by philosophical dogmas. In controversial cases, as, e.g., in the case of the AC in set theory, their attitude can be characterized as neutral – without making any philosophical declarations they simply considered and studied various mathematical consequences of both accepting and rejecting controversial principles and investigated their role in mathematics.

Note. The financial support of National Center for Science [Narodowe Centrum Nauki], grant No N N101 136940 is acknowledged.

The philosophy of Mathematics and Logic in Cracow between the Wars

The aim of this chapter is to present a philosophical reflection on the logic and mathematics in Cracow in the interwar period. This was a time of the intensive development of mathematics (Polish School of Mathematics) and of logic (Warsaw School of Logic, Lvov–Warsaw Philosophical School) in Poland. One can ask whether this development was accompanied by philosophical and methodological reflection. We shall present and analyse philosophical concepts concerning the logic and mathematics of Jan Sleszyński, Stanisław Zaremba, Zygmunt Zawirski, Witold Wilkosz and Leon Chwistek who were active (mainly) in Cracow. Their ideas will be confronted and compared with the philosophical concepts of the logicians and mathematicians who were members of the Warsaw School.

The connections of the indicated scholars with Cracow were of a varying nature and degree of intensity. There is no doubt that Sleszyński, Zaremba and Wilkosz should be associated with Cracow. The situation is no longer so simple in the case of Zawirski and Chwistek. Zawirski obtained his Habilitation at the Jagiellonian University in Cracow in 1924 and then between 1924 and 1928 he was professor at the Technical University in Lvov and then, subsequently, from 1928 until 1937 he was at the University in Poznań and from 1937, professor of the Jagiellonian University in Cracow. Therefore he is sometimes treated as a member of the Lvov–Warsaw Philosophical School. Chwistek, on the other hand, obtained his doctorate and his Habilitation in Cracow and was active there until 1930 when he became professor of the Faculty of Mathematics and Natural Sciences of the Jan Kazimierz University in Lvov. He is sometimes associated with the Lvov School of Mathematics but, nevertheless, we have decided to include Zawirski and Chwistek in the group of philosophers associated with Cracow.

Jan Sleszyński

Jan Sleszyński (1854–1931), who worked in mathematical analysis and in the calculus of variations as well as in number theory, can – in a certain sense – be treated as a pioneer of mathematical logic in Poland. He stressed the meaning and role of it for mathematics. He saw in it a tool which enabled mathematical reasoning to be made clear and precise. It is important because, as he wrote: "Everything that is obscure and complicated has no value" (cf. 1925–1929, vol. II, p. 212).

Sleszyński distinguished in mathematics between the context of discovery and the context of justification. In the first one, intuition plays an important role. Its results should be made precise and clear using methods of logic. Hence the need to construct complete proofs of mathematical theorems – they are necessary in foundational studies. "Normal" mathematicians do not usually give such proofs – they are satisfied by sketches of proofs. One should elaborate methods that will help to avoid mistakes in such reasonings. And this is the task of logic understood by Sleszyński as a theory of proof. He realized this task in his lectures devoted alternately to the methodology of mathematics and to mathematical logic. The notes to those lectures (prepared by his students) were published as a book *Teorja dowodu* [Proof Theory] (two volumes, 1925 and 1929). It contains general considerations on the concept of a deductive system understood as a collection of sentences, some of which have been assumed without proof (axioms and definitions) and others which have been deduced from them. Furthermore, there is some historical information on the development of logic and an analysis of the concept of a deductive proof in mathematics. The second volume, describing the contribution of several thinkers to the development of modern formal logic, stresses the passage from a logical calculus which serves as a tool to solve logical problems to calculus as a tool for the justification of theorems. It also contains an exposition of a system of propositional calculus and an analysis of several examples showing how rules and laws of logic can be applied to construct complete proofs of mathematical theorems.

One of the greatest merits of Sleszyński was the attempt to formulate and realize a program of the reconstruction of the real process of proving theorems in mathematics. It was later undertaken by Stanisław Jaśkowski in the form of a system of natural deduction. It should be stressed here that Sleszyński rejected psychologism, and he treated a deductive system as a hypothetical system and emphasized that mathematical theorems are in fact statements about connections between an antecedent and a consequent.

Sleszyński saw the role and meaning of symbolism in logic and mathematics but simultaneously warned against – as Kazimierz Twardowski called it – "symbolomania and pragmatofobia". Symbols serve only as a tool to make proofs simpler and more transparent.

Sleszyński – although he avoided ontological considerations – was in fact an anti-fictionalist. He treated the fictions, especially inconsistent fictions, which had been introduced into mathematics as being destructive – they usually appear at those places where the used concepts are not precise enough.

The most important contribution of Sleszyński to the development of the philosophical and methodological reflection on mathematics was his conviction of the role and meaning of logic and of formal methods for mathematics and its methodology. But one should stress that he did not treat logic as an independent and autonomous discipline – just the opposite, for him it was an auxiliary discipline with respect to mathematics. Consequently, he considered and studied it exclusively in the context of its possible applications in mathematics and not for its own sake. This view was characteristic and typical for scientists in Cracow – we shall see similar views in the case of Zaremba. This distinguished them from the Warsaw School, where logic was treated as an independent and autonomous discipline which formed the foundations and methodology of mathematics.

Stanisław Zaremba

In this way we come to the next important figure in Cracow, namely Stanisław Zaremba (1863–1942). He was interested in modern mathematical logic and, although he did not work himself in it, he saw and appreciated its role for mathematics. He was interested not in logic itself but rather in the foundations of mathematics and the analysis of the logical structure of mathematics. One can see here the influence of his studies in France (where he obtained his doctorate in 1899). His views towards logic and its role in mathematics corresponded to the attitude in this respect to that of most French mathematicians.

Zaremba was a versatile and multifaceted mathematician. He worked mainly in mathematical analysis, in particular in the theory of partial differential equations of the second order. It could be a consequence of his conviction that mathematics should not be a goal for itself, that the ultimate aim of mathematics is its applications, in particular applications in natural sciences. Nevertheless, he also appreciated the meaning and the beauty of pure mathematics.

Considering the independence of the fifth postulate of Euclid (concerning the parallels), Zaremba (1911) distinguished the provability of certain propositions on the basis of accepted axioms and their truth.

Being interested in the applications of mathematics, he considered relations between mathematics and physics (cf. Zaremba 1923, 1938). He treated this problem as very important since its solution can help us to understand both disciplines better. It is clear that mathematics is an indispensable tool for physics but also mathematics needs physics – various physical problems stimulated research in mathematics and contributed to the development and evolution of mathematics, e.g. the theory of vibrations of a string or the problems connected with the theory of heat.

Though the deductive method – which is characteristic for mathematics – is used in physics, one should stress that deductive reasoning does not justify the claim that the thesis which has been proved is true. It depends on the truth of accepted assumptions used in the proof. Additionally, it is usually the case that the postulates of any deductive theory can be interpreted in various ways. Hence those theories do not concern a particular kind of concrete objects. Mathematics provides tools for the deduction of consequences from assumed empirical hypotheses. Those consequences are then checked by experiments, and in this way one can confirm or reject those hypotheses. Hence, on the one hand, mathematics plays an auxiliary role towards physics providing it with formal methods and, on the other, it was developed (among others) by attempts to solve the problems suggested by physics.

As said above, Zaremba was interested in mathematical logic. He treated it as an auxiliary discipline with respect to classical fields of mathematics, as a tool in the didactic of mathematics. Consequently, he was not interested in the theoretical inner problems of logic itself. According to him, logic should be "in mathematics" (*en mathématique*), should be *ancilla mathematicae* as he wrote in his work *La logique des mathématiques* (1926). His views were here in full accordance with the views of French mathematicians and in contrast with the views of the members of the Warsaw school of logic.

This divergence of opinion was demonstrated in particular in the polemics between Zaremba and Jan Łukasiewicz – it was described extensively by Jan Woleński in his book *Szkoła Lwowsko-Warszawska w polemikach* [Lvov-Warsaw School in Polemics] (1997).

In 1916, Łukasiewicz devoted one of his courses at Warsaw University to the methodology of deductive sciences. During his lectures he discussed the book *Arytmetyka teoretyczna* [Theoretical Arithmetic] by Zaremba (1912) and analysed it from the methodological point of view, challenging some principles adopted by Zaremba as well as his definition of a magnitude (in particular he criticized the usage of sentences with no contents). Łukasiewicz published his remarks in a paper (1916). This was the beginning of a dispute in which several persons took part, including among others Kazimierz Kuratowski, Tadeusz Czeżowski, Leon Chwistek and, of course, Zaremba. The essence of the dispute concerned in fact not the very concept of a magnitude – this problem can be solved by relativization of formalism to a certain given interpretation or model – but the role of logic in mathematics. Zaremba was of the opinion that logic should play an auxiliary role in mathematics – it should help us to construct correct mathematical reasonings. Hence logic belongs in fact to the propaedeutics of mathematics and cannot be itself a subject of studies. Consequently, one cannot speak about a priority of logic towards mathematics. The requirements to provide complete proofs and the idea that definitions are in fact superfluous (since they can be eliminated) – as postulated by the new mathematical logic – are, according to Zaremba, only ballast and they rather hinder the process of understanding and communicating mathematical results.

The opinion of Łukasiewicz was different. He was convinced that incomplete proofs are not only didactically imperfect but they can be also a source of errors. Science is not a collection of true statements – it should be an edifice in which every element is connected with the whole. Only logic helps and enables us to uncover those connections, and logic provides tools to discover such connections. The Warsaw School of Logic shared the opinion of Łukasiewicz and treated logic as an autonomous discipline, as a discipline that belongs to the core of mathematics. This view corresponded with the program of Warsaw School of Mathematics (Sierpiński, Janiszewski and Mazurkiewicz) where stress was put on set theory, foundations of mathematics and just logic (cf. Murawski 2010).

Zygmunt Zawirski

Zygmunt Zawirski (1882–1948) was mainly interested in the philosophy of science, in particular in the methodological, epistemological and ontological problems connected with physics. He was also interested in the problem of time. He represented moderate realism, criticized neo-Kantianism and Empirio-criticism. He was also interested in the problem of applications of results of formal disciplines in natural sciences.

The philosophical problems of logic and mathematics appear rather at the margins of his considerations. One can distinguish here two circles of problems that interested him: the connections between logic and mathematics and the meaning and role of non-classical logics. The main aim of his papers in this domain was to inform the philosophical circle about new achievements in the world and to correlate them with research undertaken in Poland. Consequently, he rather rarely formulated his own opinions.

He stressed that mathematics is older than logic – the ancient Greeks formulated correct mathematical proofs long before systematic investigations on the essence of logical reasoning started. The deductive sciences investigate formal objects whereas in the natural sciences one is interested in the examination of the real world. One searches there not for *ens* (as in logic and mathematics) but *ens existens* – and this can be done only with the help of empirical methods. However, it does not mean that logic and mathematics have no meaning for natural sciences. For mathematical theories can be interpreted. In this way, mathematical constructions become the components of physical theories and mathematical theories interpreted in such a way can be checked empirically.

Zawirski was interested in particular in the connections between physics and geometry. According to a classical approach, i.e. before the development of non-Euclidean geometries – both investigate the space. Thanks to axiomatization one can claim that the differences between physics and geometry disappear: physics becomes interpreted geometry (Schlick) and geometry becomes a natural science (Einstein, Born). According to Zawirski, physics and geometry are distinct disciplines. Geometry constructs its subject independently of experience and of the existing physical reality. It justifies its theses exclusively by deduction. Physics, on the other hand, investigates objects given in experience and formulates its theses usually using inductive methods. In a developed physical theory, one can justify particular theses by deducing them from accepted axioms – this can be the case only in the context of justification. However, accepted axioms must have a certain empirical justification. In geometry, similarly as in the whole of mathematics, certain statements and laws can have empirical sources but they can be accepted only after deducing them from the axioms. Experience is no justification in mathematics.

How did Zawirski understand logic? According to him, logic is a discipline that considers forms of reasoning which are used in any inference or proof. It reflects the common structure of justification used in all disciplines. Hence Zawirski understood logic in a broad sense – not only as a formal system (or a collection of such systems) but also as a discipline about any reasoning. One should add that such opinion was rather popular in Poland at the time.

Let us come now to the second point, i.e. to the problem of non-classical logics. Zawirski was interested mainly in intuitionistic and many-valued logics. In particular, he was interested in the problem of the possible applications of many-valued logics to the solution of problems of quantum mechanics or of problems generated by the introduction to physics of statistical laws. He claimed that the new many-valued logic was the only way to help understand the phenomena of the microworld. Using some of the ideas of Łukasiewicz and Post, he attempted to develop a system of logic that would be appropriate both to the problems of modern physics as well as to probability calculus. He described the certain parallelism between expressions of the probability calculus and formulas of the many-valued logic of Łukasiewicz and Post. He claimed that both should be treated as different systems providing an empirical base for the other. Such an approach would make possible the application of many-valued logic in quantum mechanics. This idea was developed further in particular by P. Suppes. In this way, Zawirski can be treated as a precursor of quantum logic.

Witold Wilkosz

Under the influence of Sleszyński, some young mathematicians in Cracow became interested in logic – among them was Witold Wilkosz (1891–1941).

He did not write any separate papers devoted to the philosophy of logic or mathematics, nevertheless in his works one finds several remarks of a philosophical nature. They contain neither revolutionary ideas nor form a compact system, yet still they indicate tendencies and philosophical sympathies and show that Polish mathematicians were interested in philosophical problems.

Let us start with his understanding of the origin of the concept of number. In a popular work *Liczę i myślę. Jak powstała liczba* [I count and I think. How did numbers originate] (1938) he accepts the logico-set-theoretical conception (as opposed to the intuitionistic one) according to which the concept of a number and the whole of arithmetic comes from more general intuitions connected with equipollency. He justifies this thesis with reference to many examples of a psychological and ethnological nature.

Logic is – according to Wilkosz – only a tool for developing a system on the basis of accepted axioms and definitions. Hence it is connected with deductive systems. Consequently, it can be successfully applied in mathematics; but attempts to apply it in other disciplines have been rather ineffective. The reason is that the axiomatic methods can be applied in situations when the starting and fundamental postulates are constant and fixed. But this is not the case in many disciplines. On the other hand, he was of the opinion that strict methods based on the application of formal logic should also be used in the humanities and in the social sciences as well as even in theology.[133]

Formal logic does not fully fit "natural" methods and the ways of reasoning used, e.g. by mathematicians. In fact, it is suitable rather to the reconstruction of a real mathematical reasonings, hence in the context of justification and not directly in the context of discovery.

Wilkosz also tried to apply formal logical methods to explain some philosophical problems, in particular the problem of abstraction. The problem is: What is the abstract object? What should it be identified with and how can it be best described? In trying to answer this, Wilkosz used equivalence relations and equivalence classes. He suggested applying the method of representation and to identify the abstract either with the appropriate equivalence class or with one element of such a class. But he also saw another solution: instead of deciding what is the abstract object, one can explain what sentences about it in fact mean.

Leon Chwistek

The most important and best-known figure among the Cracow logicians and mathematicians interested in the philosophy of logic and mathematics was Leon Chwistek (1884–1944).[134] He is known mainly for his logical works, in particular for his simplification of Whitehead and Russell's theory of types. His logical investigations, however, were connected with his philosophical ideas concerning logic

---

133 In fact such attempts have been undertaken, e.g. by the so-called Cracow Circle whose members were I.M. Bocheński, F. Drewnowski and J. Salamucha as well as B. Sobociński – they tried to apply logico-axiomatic methods to philosophical and theological problems.
134 For Chwistek's philosophy of logic and mathematics see also Murawski (2011b).

and mathematics.[135] They were motivated by those ideas and, in this sense, he was an exception among Polish logicians and mathematicians. His aim was not only to solve particular fragmentary problems, but (as it was also the case by Leśniewski) he attempted to construct a system containing the whole of mathematics.

Chwistek (1924, 1925) formulated a pure theory of logical types – a theory of constructive types. In this theory, the nonconstructive objects are rejected but the price for that is the greater formal complication of the system.

Those investigations led Chwistek to the construction of a full theory of expressions and of the so-called rational metamathematics. The latter should be a system more fundamental than logic. It should enable the reconstruction of a classical logical calculus and of Cantor's set theory. According to Chwistek, it should be based on nominalistic assumptions – hence in particular it should be free of any existential axioms, first of all of the reduction axiom and the AC. The basic assumption was that theorems of the system being constructed, and consequently theorems of classical logic and of set theory, refer only to expressions/inscriptions that can be obtained in a finite number of steps by a rule of construction fixed ahead and not to the meaning of those expressions. Moreover, those expressions/inscriptions were understood as physical objects.

All of those ideas were developed by Chwistek later as part of his system of philosophy of logic and mathematics, in particular as a part of his methodology of deductive sciences. He developed it mainly in the book *Granice nauki. Zarys logiki i metodologii nauk ścisłych* from 1935 – the English translation *The Limits of Science. Outline of Logic and of the Methodology of the Exact Science* appeared in 1948.

According to Chwistek, human knowledge is neither complete nor absolute. It cannot be complete because statements concerning the totality of objects lead to inconsistencies. On the other hand, it cannot be absolute because there exists no absolute reality. In *The Limits of Science* he wrote (1948, p. 42):

> It follows from these considerations that the principle of contradiction does not permit complete knowledge, i.e., knowledge which includes the answer to all questions. The attempts to secure such knowledge will sooner or later conflict with sound reason.

And common sense is – according to Chwistek – a factor common to all correct cognitive processes. It functions beside the admission of experience as a fundamental source of knowledge and of the necessity of schematization of experienced objects and phenomena. Common sense consists in rejecting all assumptions that cannot be experimentally checked or are inconsistent with experiments or are not based

---

135 Similar interconnections between philosophical ideas and research in logic one can see also in the case of Stanisław Leśniewski (1886–1939) – cf. Murawski (2004a and 2011a as well as 2014).

on reliable and certain statements concerning simple facts or cannot be logically reduced to such statements. Both empirical and deductive knowledge are relative. Empirical knowledge is relative because there are various types of experiments corresponding to various realities, and the deductive one – because it depends on the accepted system of concepts. Chwistek is talking here about rational relativism.

Chwistek was decidedly against irrationalism – he accepted the principle of the rationalism of knowledge. Rationalism consists in accepting only two sources of knowledge, namely experience and strict reasoning. It concerns not only mathematics and the exact sciences but experimental sciences and philosophy as well. Chwistek wrote in *The Limits of Science*: "[...] the point of departure in constructing a world view should be not a confused metaphysics, but simple and clear truths based upon experience and exact reasoning" (1948, p. 3).

Consequently, he was against irrationalism, metaphysics and idealism in philosophy and mathematics.[136] He sharply criticized Plato, Hegel, Husserl and Bergson. Despite the defects of positivism, he appreciated its epistemological conceptions. He also greatly appreciated dialectical materialism, seeing in fact almost no fundamental conflict between it and positivism. His own epistemological conceptions he described as critical rationalism and set it against dogmatic rationalism.[137]

A way out of the difficulties caused by irrationalism and simultaneously a weapon in a struggle against it is formal logic, in particular the rational metamathematics founded by him. Chwistek began his book *The Limits of Sciences* by writing in the first sentence (cf. 1948, p. 1): "We are living in a period of unparalleled growth of anti-rationalism". And he finishes the Introduction by writing (cf. 1948, p. 23): "History teaches that ultimately victory has always been the destiny of societies who employ the principles of exact reasoning".

In the Introduction he also wrote (1948, p. 22):

> When this new system [that is, the system of rational metamathematics – my remark, R.M.] is completely worked out, we will be able to say, that we have at our disposal an infallible apparatus which sets off exact thought from other forms of thought.

The epistemological views of Chwistek were closely connected with those of neo-positivism. He claimed that an object of a scientific knowledge can only be what is or can be given in experience, hence only what can be seen or experienced by the

---

136 It is worth noting here that Chwistek was against irrationalism and idealism not only because they are – in his opinion – incorrect philosophical theories but also because they are the source of human sufferings, social injustice, cruel excesses and wars.

137 A certain difficulty in interpreting Chwistek's philosophical views should be noted. In fact, he often used classical philosophical notions but gave them a special meaning which he never explained or explained in an insufficient way.

senses, eventually assisted by instruments. He wrote in Chwistek (1948): "Talking about reality we do not think about an ideal object but about those schemes we have to do with in a given case".

Chwistek appreciated constructive methods. He was of the opinion that they should be used both in science and in the philosophy. This method was explained by him in the paper „Zastosowanie metody konstrukcyjnej do teorii poznania" [Application of a Constructive Method to Epistemology] (1923). According to him, it can be applied in a full form mainly in deductive sciences but nevertheless it can be used in empirical sciences and in philosophy as well. The method is based on the analysis of the intuitive concepts used in a given discipline. This permits the separation of the primitive notions characterized in axioms. On the basis of axioms, new theorems can be obtained by laws of (formal) logic. Later Chwistek came to the conclusion that attempts to construct deductive systems in philosophy are in fact useless – the reason is the fact that philosophical investigations are too complicated.

As indicated above, Chwistek claimed that an object of cognition can only be what is given in an experience. There are, however, various types of experience. In this way, we come to the best-known original philosophical conception of Chwistek, namely to his theory of the plurality of realities. It was explained by him for the first time in the paper „Trzy odczyty odnoszące się do pojęcia istnienia" [Three Lectures Concerning the Concept of Existence] (1917). Chwistek (1917, p. 145) claimed there that: "the intuitive belief in one reality seems to be a superstition" and saw the concept of many realities already by Pascal and Mach (cf. 1917, pp. 149–50). The conception of the plurality of realities was developed in his book *Wielość rzeczywistości* [Plurality of Realities] (1921). The final version can be found in *Granice nauki* (1935). Its foundations were explained once again in the English version of this book (1948) but this in fact brought nothing new.

In the first period of his scientific activity, that is up until 1925, Chwistek distinguished between the concepts "reality" and "existence". The latter has – according to him – a more general character because it can concern not only objects of reality but also abstract objects such as the objects of mathematics. He wrote: "If we assumed that everything that exists is in fact real, then we should accept as real all mathematical relations together with elements of experience" (1917, p. 145).

Chwistek (1917) distinguished three possible positions concerning the problem of existence: nominalism, realism and hyper-realism. According to him "nominalists demand descriptions by words excluding inconsistencies" (1917, p. 126), realists do not demand descriptions by words but they "exclude inconsistent objects" (1917) and "hyper-realists do without descriptions by words and do not exclude inconsistent objects" (1917).

In the beginning, he accepted only two types of reality and attempted to formalize his theory. In the book *Granice nauki*, he resigned from his attempts of formalization and accepted four types of reality corresponding to possible types of experience. Hence he distinguished the reality of impressions, the reality of images, the reality of things (the reality of everyday life) and physical reality (that which is constructed in the exact sciences). Independent existence and full theoretical equality of rights were attributed to all particular kinds of reality.

Let us turn now to Chwistek's views connected directly with the philosophy of mathematics. The characteristic feature of them is nominalism.

Chwistek claimed that the object of deductive sciences, hence in particular of mathematics, are not any abstract ideal entities but in fact expressions being constructed according to accepted rules of construction. Hence the objects of mathematics are not points, lines, numbers, sets and so on, but expressions being physical objects given to us in experience. Those expressions can be transformed according to accepted rules – in every given system such rules as well as some expressions that play the role of axioms and form the basis on which one deduces theorems are accepted. The rules of transformation and axioms are chosen in such a way that the expressions can be interpreted as descriptions of considered states of things. To apply deductive theories to particular disciplines and generally to get to know particular domains of the reality, elements of the latter should be schematized.

Chwistek claimed that geometry is an experimental discipline. In Chapter VIII of *The Limits of Science* he wrote (1948, p. 170):

> Geometry is an experimental science. It depends upon the measurement of segments, angles and areas. The Egyptians conceived it in this way and it has remained essentially the same up to this very day. Today what is generally regarded as geometry, i.e., what is included in textbooks, is the peculiar mixture of experimental geometry and the geometrical metaphysics which was inherited from the Greeks as Euclid's *Elements*.

In Chwistek's opinion, the development of systems of non-Euclidean geometry of Bolyai, Gauss and Lobachevsky[138] in the 19th century rejected Kantian idealism. Those geometries have shown that, e.g. the concept of a line has no objective character but depends on adopted axioms. This can suggest that conventionalism should be a proper philosophy for geometry. Indeed, in his first papers, for example in the paper „Trzy odczyty odnoszące się do pojęcia istnienia" [Three Lectures Concerning the Concept of Existence] (1917) Chwistek stated explicitly that the existence of consistent systems of non-Euclidean geometries refutes the thesis about

---

[138] It is worth noting that Chwistek considered them to be the most important achievement in the exact sciences.

the character of geometry. Nevertheless he never stated explicitly that he would be ready to accept conventionalism. He wrote (1917, pp. 144–145):

> Both systems [that is, the system of Euclidean geometry and systems of non-Euclidean geometries – my remark, R.M.] are free of inconsistencies – in fact they can be reduced to analytic geometry. Hence there are almost no fundamental differences between them from the theoretical point of view. Intuition easily accepts Lobachevsky's theorems that only at first glance seem to be paradoxical [...]. So we come to the conclusion that both geometries are equally true, each of them refers to different lines; the differences between those two types of lines can be caught neither by experimental means nor by intuitive ones, hence a segment of a line we draw or think can serve as an illustration of one or another type depending on our will.

In *The Limits of Science*, however, Chwistek clearly and categorically rejected conventionalism. He claimed that geometry – similarly to all other fundamental experimental sciences – should be based on a theory of expressions. The reason is that in fact conventionalism introduces hypothetical entities (it was the case already in J.S. Mill or later by Poincaré, the propagator of this tendency).[139] He wrote (1948, pp. 186–187):

> It seems that it is impossible to attain the general concept of a geometry without using formulae. It is therefore clear that the conception of geometry as the science of ideal spacial constructions must be nullified. [...] To speak of different four-dimensional space-time it is necessary to employ five-dimensional space-time. It is clear that all this has only as much meaning as do mathematical formulae.

One should also treat arithmetic, mathematical analysis and other mathematical theories in a similar way as geometry, obtaining in this way a nominalistic interpretation of all of them.

To sum up, add that philosophical investigations of Chwistek had no systematic character; and it seems that they were not treated by himself with a full sense of responsibility (cf. Chwistek 1961, Preface, p. vii). He did not explain many of the concepts he used; his conceptions has been "earlier proclaimed than checked" (Chwistek 1961). He did not develop his systems in detail but satisfied himself by sketching them only. Chwistek's system of rational metamathematics could neither have been developed by his collaborators (among them were Jan Herzberg, Władysław Hetper and Jan Skarżyński) nor by his students and pupils (Wolf Ascherdorf, Celina Gildner, Kamila Kopelman, Abraham Melamid, Józef Pepis and

---

139 It is worth adding here that Chwistek rejected conventionalism not only for this reason. He claimed also that conventionalism became a source of reactionary social views and tendencies by reducing truth to efficiency and leading in this way to the reinforcement of ruling classes. In 1948 (p. 234) he wrote: "It should be observed that idealism clothed in the feathers of conventionalism became a very dangerous instrument in the hands of those who were reacting against the old dogmatic idealism".

Kamila Waltuch) because all of them were killed during the Second World War. His investigations were not in the main stream of the development of logic and philosophy of mathematics. Therefore his works did not generally find any interest amongst logicians and philosophers (with the exception of his version of type theory). Similar to Leśniewski (cf. Murawski 2004a), Chwistek worked on his own conceptions and ideas without any collaboration with other logicians, mathematicians or philosophers; he went solitarily along his own paths. Being a professor of Lvov University, he had in fact no stronger contacts with the Lvov–Warsaw Philosophical School (cf. Woleński 1989). Only after 1945, together with the growing interest in the nominalism in the philosophy of mathematics, did some of his ideas find recognition. One should also add that recently a reference to Chwistek's pluralism was made by the Australian philosopher R. Sylvan (1997).

Conclusion

As we see mathematicians, logicians and philosophers in Cracow (or connected with Cracow) were interested in the philosophical problems of logic and mathematics but rather on the margins of their proper scientific work. The sole exception here was Chwistek, whose logical investigations were connected with his philosophical ideas concerning logic and mathematics and were in a sense motivated by them. What is characteristic for all of the Cracow scientists presented here is that they were interested in logic not as an autonomous discipline but treated it only as a useful tool that could help us to make the mathematical reasoning more precise and clear. According to them, logic should provide practical tools for constructing proofs and justifying reasonings; it is an auxiliary discipline and can be reduced to a theory of proof or rather a theory of constructing (first of all mathematical) proofs. This is the difference with respect to the Warsaw School where logic was developed as an independent discipline. In fact, the approach to mathematics represented by Cracow mathematicians was rather classical and not modern – as it was the case in Warsaw or Lvov. This was rightly characterized by the Russian mathematician Nikolai Lusin who in a letter to Arnold Denjoy wrote in 1926 (cf. Lusin 1983, p. 66):

> It seems that the mathematical life in Poland is being developed along two different paths: one of them leans toward classical fields of mathematics, the other – toward set theory (theory of functions). Those tendencies in Poland exclude each other, each of them is hostile toward the other and a stout battle between them goes on. Both parties are very vigorous but, as seems to me, their strengths are not equal. [...] The classical side is represented nowadays only by the old [...] Cracow University. [...] The most inflexible follower of this tendency among Polish mathematicians is professor Zaremba. Other adherents are close to him. [...] However the classical tendency came to the end in many cities [...] where it has been replaced by the tendency of the school of Mr Sierpiński.

Note. The financial support of the National Center for Science [Narodowe Centrum Nauki] (Grant No N N101 136940) is acknowledged.

Philosophy of Logic and Mathematics in the Lvov School of Mathematics

The aim of this chapter is to consider philosophical ideas concerning logic and mathematics developed in Lvov School of Mathematics. Views of Hugo Steinhaus (1887–1972), Stefan Banach (1892–1945), Eustachy Żyliński (1889–1954) and Leon Chwistek (1884–1944) will be analysed. In the case of the first three of them, there is no room for doubt that they belonged to this school. There may be some doubts in the case of Chwistek. We have included him into the Lvov school because since 1930 he was the chairman of the chair of mathematical logic at the faculty of mathematics and natural sciences of the Jan Kazimierz University in Lvov – though some part of his scientific career was connected with Cracow, he developed his main philosophical ideas just in Lvov.

Lvov School of Mathematics, accepting main ideas of Janiszewski's program (1917), developed another specialization than the Warsaw school. In Warsaw, mainly set theory, topology and mathematical logic were developed. In Lvov, functional analysis dominated, which was initiated by Stefan Banach (his mathematical talent has been discovered by Steinhaus) and developed by Steinhaus, Stanisław Mazur, Władysław Orlicz, Juliusz Schauder, Stefan Kaczmarz, Stanisław Ulam and Władysław Nikliborc. It did not demand deeper studies of logic and foundations of mathematics as it was the case in Warsaw. Consequently, it is rather difficult to find philosophical remarks concerning mathematics in works of Lvov mathematicians. It could be also the result of the fact that logic as such has not been developed in Lvov, though the intellectual atmosphere for it and for the foundations of mathematics was good here (cf. Woleński 2009). Only in 1928 it has been decided to form a chair for mathematical logic – its first chairman became Chwistek. Earlier the only Lvov mathematician who worked in logic was Eustachy Żyliński. One should add, however, that other Lvov mathematicians did not disparage logic and the foundations of mathematics or even casually worked in it – one should mention here Banach and his joint paper with Alfred Tarski on the paradoxical decomposition of sphere (cf. Banach, Tarski 1924) or results of Banach and Mazur concerning the computational analysis and constructive methods in mathematics (cf. Banch, Mazur 1963).

Stefan Banach

Stefan Banach did not avoid to take part in the philosophical life of Lvov and from time to time was active there. In particular Kazimierz Twardowski in his *Dzienniki* [Diary] (1997, vol. 1, p. 201) writes that Banach took part (on 7th March 1921) in the inaugural meeting of the Section of Epistemology of Polish Philosophical Society and that he was present at the talk by Zygmunt Zawirski on relations between logic and mathematics held on 26th March 1927 during a meeting of Polish Philosophical Society (*ibidem*, p. 300). On the 1st Congress of Polish Mathematicians held in Lvov in 1927, Banach gave (on 7th September 1927) in the section of mathematical logic a talk „O pojęciu granicy" [On the concept of a limit] (cf. Twardowski 1997, vol. 1, p. 323). In January 1923 at the meeting of Polish Philosophical Society in Lvov, Banach gave a talk on paradoxes connected with the concept of equipollence of certain sets (e.g. the set of integers and the set of even natural numbers) as well as on problems connected with Banach-Tarski paradox. As source of those paradoxes he indicated infinite sets and the AC (formally consistent with set theory). According to him, a logical system that "would not awake any objections" should be constructed to solve those paradoxes. This remark characterizes the attitude of Lvov mathematicians towards logic. In particular, Banach did not see anything wrong for the mathematical practice in the lack of a good logical system. In the Lvov School of Mathematics the development of mathematics did not require additional studies in logic and the foundations of mathematics.

Hugo Steinhaus

The best way to reconstruct the picture of mathematics cherished in Lvov is to analyse some remarks contained in works of popular character, in particular in works by Steinhaus.

One should tell here first of all about his book *Czem jest a czem nie jest matematyka* [What is and what is not mathematics] (1923). He writes there about various topics, in particular about the definition of mathematics, about its historical development, practical applications, method of mathematics, about differential and integral calculus, about numerical mathematics, about errors in mathematics and about connections of mathematics with the everyday life. From our point of view, the most important are his remarks on defining mathematics as a science and his considerations about mathematical methods.

Trying to define mathematics, Steinhaus stresses that on the one hand mathematics grew out from some practical needs of human being but on the other it is in fact a theoretical discipline. A characteristic feature of mathematics is its deductive method. He adds that "its axioms and definitions are in a large extent arbitrary" (1923, p. 25). Another feature of mathematics is the usage of symbols.

Logic is treated by Steinhaus with sympathy but not as an independent discipline having its own problems and methods. He treats logic as a tool of deduction. The deductive method determines in a certain sense also the subject of mathematics.

Mathematics is deductive, synthetic and formal. It is deductive since deduction is the only method allowed in it. It is synthetic because axioms, both logical and mathematical, are chosen not logically but with the help of intuition. It is formal because in mathematical argumentation one can take into account only those elements of concepts that have been included in definitions. Logic plays only an utilitarian role towards mathematics providing it with tools.

In the development of mathematics, an important role is played also – according to Steinhaus – by aesthetical elements. Though there are no absolute criteria of beauty, in fact the feeling of beauty and the aspiration for it influence more the development of mathematics than the principle of perfect precision.

Steinhaus appreciated very much applied mathematics and applications of mathematics. Unfortunately, he did not describe the connections between concepts and objects of mathematics on the one side and the reality on the other. One finds only his short and aphoristical remark: „Między duchem a materią pośredniczy matematyka" [Between spirit and matter mediates mathematics].

Eustachy Żyliński

Eustachy Żyliński worked mainly in number theory, but after 1919 he began to work in algebra, logic and the foundations of mathematics. In particular he proved (por. Żyliński 1925; see also 1927) that in the classical propositional logic the only functors that suffice to define all other functors are binegation and Sheffer's disjunction.[140] One finds no separate papers by Żyliński devoted to the philosophy of mathematics and logic. We have only some remarks of philosophical character he made on various occasions.

In a talk (held on 21st May 1921) „O przedmiocie i metodach matematyki współczesnej" [On the Subject and Method of Contemporary Mathematics] he identified mathematical theories with the set of consequences of accepted axioms. One should note here his unprecise treatment of logic – Żyliński refers to subjective feeling of certainty and obviousness rather than to formally and in advance described inference rules. He admits an infinite set of consequences of accepted axioms talking about corollaries that can be obtained.

Considering the problem of relations between logic and mathematics, he compares it with the relation between "special set theories and a general one". He claims that mathematics is a natural science about certain objects and says that one refers to observation and even experiment in developing particular mathematical theories.

In the paper „Z zagadnień matematyki. II. O podstawach matematyki" [Problems of Mathematics. II. On Foundations of Mathematics] (1928) Żyliński says about intuition. He argues that intuition can help to construct a proof but stresses

---

[140] A proof of this theorem can be found in Murawski, Świrydowicz (2006).

that the proof itself cannot refer to intuition. Hence one has here the distinction between the context of discovery and the context of justification. In the first one – intuition is admitted, in the second – not.

He saw the great role played by mathematics in other disciplines as well as generally in culture. In a memorial by Żyliński, Ruziewicz and Banach from 14th April 1924 one reads (Żyliński et al. 1924, p. 1):

> Contemporary mathematics is nothing else as a general theory of strict thinking connected with the feeling of certainty. [...] Being a most general science about relations between objects, mathematics finds applications in every scientific and practical discipline that goes out in a sufficient manner beyond a description, simple induction and literary-artistic methods.[141]

[...][142]

Conclusion

The above considerations show that in the Lvov School of Mathematics no general, comprehensive and homogeneous philosophical concept concerning mathematics and logic has been formulated and developed – the only exception was here Chwistek. One finds there only separate, detached remarks formulated when considering other problems and being in fact a reflection on own research in mathematics. Only Chwistek who was in fact not a mathematician but a (mathematical) logician tried to develop certain comprehensive theory. The dominating feature of his approach were nominalism and constructivism with all their consequences. Unfortunately, the style of his work did not allow to develop his concept with all details.

Note. The financial support of National Center for Science [Narodowe Centrum Nauki], grant No N N101 136940 is acknowledged.

---

141 Matematyka dzisiejsza jest niczym innym jak ogólną teorią ścisłego myślenia połączonego z poczuciem pewności. [...] Będąc jednak najogólniejszą nauką o relacjach zachodzących między przedmiotami, matematyka znajduje zastosowania w każdej dziedzinie naukowej i praktycznej, wychodząc w dostatecznej mierze poza ramy opisowości, prostych indukcji lub metod literacko-artystycznych.

142 To avoid repetitions, we omit here the section devoted to Leon Chwistek – we wrote about him in the chapter "Philosophy of Mathematics and Logic in Cracow between the Wars" included into this volume.

Cracow Circle and Its Philosophy of Logic and Mathematics

1. The term "Cracow Circle" is used to describe a group of scientists who tried to apply the methods of modern formal/mathematical logic to philosophical and theological problems, in particular they attempted to modernize the contemporary Thomism (the trend which was then prevailing) by the logical tools. The group consisted of: the Dominican Father Józef (Innocenty) M. Bocheński,[143] Rev. Jan Salamucha,[144] Jan Franciszek Drewnowski[145] and the logician Bolesław

---

[143] Józef (Innocenty) Maria Bocheński was born in Czuszów (District of Miechów) on 30 August 1902. He began the studies of law in Lvov in 1920. Then he moved to Poznań (in 1922) where he studied economy. However, he did not finish his studies. In 1926, he entered the major theological seminary in Poznań, and in 1927, he joined the Dominicans. In the years 1928–1931, he studied philosophy in Fribourg (Switzerland), where he received his doctor's degree. In the years 1931–1934, he studied theology at the Angelicum in Rome – he also obtained his doctorate there. From 1934, he lectured on logic at the Angelicum. In 1938, he presented his *Habilitationsschrift* to the Theological Faculty of the Jagiellonian University in Cracow. During the Second World War he served as chaplain to the Polish Army and the Polish II Corps in Italy. In the years 1945–1972, he was a professor at the University of Fribourg. In 1948, he received the Chair of the History of Modern Philosophy there. In 1958, he founded the Institute of Eastern Europe at the University of Fribourg. He died in Fribourg on 8 February 1995.

[144] Jan Salamucha was born in Warsaw on 10 June 1903. In 1919, he entered the major seminary in Warsaw, and in 1925 he received the Sacrament of Holy Orders. In 1920, he was a medical orderly in the Polish-Soviet war. In the years 1923–1927, he studied philosophy, mathematics and mathematical logic at the University of Warsaw where he listened to the lectures by Jan Łukasiewicz, Stanisław Leśniewski, Tadeusz Kotarbiński, Władysław Tatarkiewicz and Stefan Mazurkiewicz. He obtained his doctor's degree in 1927. From 1927 till 1929, he studied at the Pontifical Gregorian University in Rome. Till 1933, he lectured on philosophy at the major seminary in Warsaw. In 1933, he completed his *Habilitation* but it was recognized by the ministry only in 1936. In the year 1934, he also began lecturing at the Jagiellonian University. In November 1939, he was arrested, together with other professors of the Jagiellonian University, and transported to the concentration camp in Sachsenhausen and from there to Dachau. He was released in January 1941 and forced his way to Warsaw where he was active in the underground movement. During the Warsaw Uprising he was a chaplain. He was killed on 11 August 1944.

[145] Jan Franciszek Drewnowski was born in Moscow on 2 December 1896. He lived in Warsaw from 1903. In 1914, he attended technical courses in Warsaw and in 1915 he studied at the Mathematical-Physical Faculty in Petersburg. In 1916, he completed a course of engineering at the Military School in Petersburg and served as officer in the Russian

Sobociński who collaborated with them. 26 August 1936 is regarded as the foundation date of the Cracow Circle.[146] On that day a special meeting was held during the Third Philosophical Congress in Cracow. The meeting gathered 32 people, including professors of philosophy of the theological academies and major theological seminaries as well as the future members of the Circle. It was presided over by the outstanding philosopher and specialist in Medieval studies Rev. Konstanty Michalski. Another participant was Jan Łukasiewicz, one of the key representatives of the Lvov–Warsaw Philosophical School[147] – specifically of the Warsaw School of Logic – who himself had dealt with philosophy and formulated a program of a radical reform of this domain, suggesting the use of the methods of modern logic. Łukasiewicz formulated this program in the paper „O metodę w filozofii" [On Method in Philosophy] (1927). During the meeting Łukasiewicz, Bocheński, Salamucha and Drewnowski presented their views and then a discussion was held. The proceedings were published in 1937 in volume 15 of *Studia Gnesnensia* under the title *Myśl katolicka wobec logiki współczesnej* [Catholic Thought in Relation to Modern Logic].

In fact, the contacts and collaboration between those who were members of the Cracow Circle began earlier – cf. Bocheński (1989) – and the above-mentioned meeting was rather a public manifestation. According to Bocheński, the Circle existed for 7 years – from the beginning of his friendship with Salamucha till the outbreak of the Second World War.

The four people who composed the nucleus of the Circle shared interests in mathematical logic as well as philosophical and theological issues. Bocheński was a doctor of philosophy and theology; he was a professor at the Angelicum in Rome.

---

army. In 1918, he was drafted into the Polish army and in 1920 he was an officer of the general headquarters. At the same time, he attended lectures at the Faculty of Finance and Economics of the School of Political Sciences. In 1921–1927, he studied philosophy, mathematical logic and mathematics at the University of Warsaw under the supervision of Stanisław Leśniewski, Jan Łukasiewicz and Tadeusz Kotarbiński. In 1927, he obtained his doctor's degree under Kotarbiński's supervision. His dissertation concerned Bolzano's theses on logic. During the defence of Warsaw in 1939 he was an aide-de-camp of the commander of sappers. After the capitulation he was in a German prisoner-of-war camp until 1945. Then he was in the Polish Forces in Rome and England. He returned to Poland in 1947. He was an advisor to the minister in the Central Planning Office and a scientific consultant in the Institute of Economics and Organisation of Industry. He was also an editor of technical dictionaries. He died in Warsaw on 6 July 1978.

146 Here we cannot give more details about the history of the Cracow Circle – more information on this theme can be found, e.g. in Wolak (1993, 1996), cf. Bocheński (1989) as well as Woleński (2003).

147 On the Lvov–Warsaw Philosophical School see Woleński (1989).

Salamucha studied philosophy, mathematics and mathematical logic at the University of Warsaw, received his PhD in philosophy at the Jagiellonian University, studied at the Gregorian University in Rome and when the Circle was created he was a professor of philosophy at the Warsaw major seminary. Drewnowski, who was T. Kotarbiński's disciple, was the editor and publisher of *Rocznik Handlu i Przemysłu* [Yearly Reports on Trade and Industry] in Warsaw. Sobociński, a philosopher and logician, was an assistant at the University of Warsaw, and he dealt mainly with formal logic. As opposed to the first three men, he did not publish any works on philosophy but he participated in all of the meetings of the Circle and in a way was an expert on logical problems.

The members of the Cracow Circle were fascinated with modern formal logic but were dissatisfied with the level and way of cultivating philosophical and theological reflections of their times. Consequently, they proposed a complete axiomatization and formalization of the Catholic doctrine, especially Thomism. It should be added that they all respected Thomism. Salamucha and Bocheński regarded themselves as Thomists. Nevertheless, they wanted to change it and transform it into a normal scientific theory. They thought that Catholic thinkers were not faithful to their sources, i.e. scholasticism. Rejecting modern logic they did not follow the spirit of St Thomas Aquinas who had made use of then existing logical apparatus in his philosophical and theological analyses. The Circle postulated a reform of philosophy, first of all its methods and not its content. They did not intend to give up traditionalistic philosophy but wanted to make it precise and develop it in a scientific way. Moreover, the representatives of the Circle thought that the new research methods, using the instruments of modern logic, allowed them to discover numerous valuable elements in the old philosophical and theological texts. They were highly critical about the philosophical systems that had originated between the sixteenth and the 19th centuries, including neo-Scholasticism. Their criticism focused on the methodology of those systems. In particular, they criticized Hegel's philosophy "not because it was idealistic, but because it was confused, badly stated and insufficiently justified" (Bocheński 1989, p. 12). Additionally, the Circle rejected neo-positivism and all minimalistic philosophies.

As mentioned before, the members of the Cracow Circle were predominantly concerned with methodological problems. They aimed at reforming the traditional way of thinking and writing, which characterized Catholic philosophers and theologians. In addition, they were convinced that modern mathematical logic could be used in philosophical and theological investigations. As Bocheński (1989, pp. 11–12) writes, they postulated that

> (1) the language of philosophers and theologians should exhibit the same standard of clarity and precision as the language of science; (2) in their scholarly practice they should replace scholastic concepts by new notions now in use by logicians, semioticians, and

methodologists; (3) they should not shun occasional use of symbolic language. To put it briefly the Circle wanted to persuade catholic thinkers and writers to adopt the "style" of philosophizing cultivated by the Polish logical school.

Łukasiewicz's influence on the Circle and its program was obvious. Bocheński writes (1989, p. 12):

> This is not surprising as all the members of the Circle, with the exception of myself, had been his pupils. His were the methodological postulates, the criticism of modern philosophy, and the doctrine of the neutrality of logic, stated explicitly for the first time at a meeting of the Circle in 1934. And again, the inquiries by some members of the Circle into the ancient and medieval logic were in fact the continuation of the pioneering work done by Łukasiewicz.

It should be added that the Circle had to face aversion and misunderstanding shown by the followers of the official theology. Its method, using mathematical logic, aroused resistance and opposition. The philosophical interpretations formulated by means of this method were accused of anti-metaphysicism, atheism, conventionalism, relativism, pragmatism, positivism and other opposing views to Christian doctrine. The use of logical methods was connected – completely unjustifiably – with the attitude towards religion of such logicians as B. Russell, T. Kotarbiński or the whole Vienna Circle. Refuting these accusations, the representatives of the Cracow Circle firmly defended the neutrality of mathematical logic with respect to philosophy. Thus they shared the views of the Lvov–Warsaw School, opposing the Vienna Circle.

The originality and significance of the conceptions formulated by the Cracow Circle should be stressed. Later, similar attempts were made by individual scientists, e.g. Bendiek (1949, 1956) or Clark (1952). However, they worked on their own account and did not form any group; consequently, their achievements are not as remarkable as those of the Circle. Bocheński thinks that the efforts of the Circle aiming at changing the Catholic thinkers' attitude towards modern formal logic were completely unsuccessful (cf. Bocheński 1989, p. 14). One of the reasons was the tragic death of Salamucha – the soul of the Circle – during the Second World War. However, the reasons were more complex. Bocheński writes (1989, pp. 15–16):

> The failure of the programme proposed by the Cracow Circle is not due to some peculiar Polish circumstances. It seems to be the result of the widespread resistance on the part of otherwise rationally thinking philosophers and theologians to recognize the significance of mathematical logic and analytic philosophy in any intellectual endeavour.
>
> The case of the Cracow Circle is particularly sad. For Poland is one of among not so many countries that has had a flourishing school of logic and an efficient team of catholic scholars, who claimed to be rational. One would have expected that in such a country a new catholic philosophy and, in the first place, a new catholic theology should arise. Alas, this has not been the case.

Nonetheless, despite Bocheński's opinions the efforts of the Cracow Circle were continued – but not as extensively as one may expect. Besides the aforementioned works of Bendiek and Clark, it is worth recalling the analyses of the five ways of Thomas Aquinas with the help of the instruments of modern logic undertaken by F. Rivetti Barbò, E. Nieznański or K. Policki.[148] These authors used the strong tools of logic like the Kuratowski-Zorn lemma (Policki) or the theory of lattices (Nieznański).

Before proceeding to analysing the philosophical views of the Cracow Circle on logic and mathematics, we should mention their main achievements as far as the implementation of the tools of mathematical logic to solve philosophical and theological problems is concerned. These achievements include:

(1) Logical analysis of the proof *ex motu* for the existence of God, presented by St Thomas Aquinas in his *Summa contra gentiles*, undertaken by Salamucha (1934),
(2) Formalisation and logical analysis of the proof for the immortality of the soul given by St Thomas Aquinas, formulated by Bocheński (1938),
(3) Analysis of the scholastic concept of analogy – these investigations were initiated by Drewnowski (1934) and Salamucha (1937a), then developed by Bocheński (1948),[149]
(4) A certain number of works concerning the history of logic, particularly the history of Medieval logic – these investigations were characterized by looking at the old logic through the prism of modern logic[150] – the works of Salamucha (1935, 1937b) or Bocheński's monograph (1956a), which was to some extent the culmination of this research trend, and
(5) Numerous works popularizing Christian thought and the new style of its cultivation.

2. Our reflection on the philosophical views on logic and mathematics formulated by the scholars under consideration should begin with the analysis of Salamucha's views. Using the methods of logic to analyse the arguments of St Thomas Aquinas, Salamucha utilized the classical two-valued propositional calculus as well as the concepts of membership, relation and set. He referred to *Principia Mathematica* by Whitehead and Russell; he also used the symbols of their work. So he neither made use of semantic concepts nor the concept of truth, nor referred to the fundamental work of Tarski (1933). The aforementioned instruments were sufficient

---

148 The analysis of these attempts can be found in Nieznański's work (1987).
149 For studies on analogy in the Cracow Circle see Wolak (2005).
150 This method was also used by Łukasiewicz – cf. Łukasiewicz (1951).

for him. Formulating his conception of logic he cut himself off from nominalism, preserving neutrality towards the philosophical problems related to his idea. In footnote 4 to his work of 1934, he wrote (cf. 2003):

> Although this way I am adopting much from mathematical logicians, it does not mean at all that I sympathize with their nominalistic point of view in logic and materialistic or positivistic tendencies in philosophy. I think that the same way as within traditional logic grounds different philosophical systems could occur equally in agreement or disagreement, it happens similarly within mathematical logic grounds, only in the second case more responsibility is required.[151]

On the one hand, he understood logic – according to Koj (1995) – as an objective science the theses of which were formulated in an objective language and not in the metalanguage. On the other hand, he treated logic as a formal science and as such, it could not be placed on any floor of abstraction. Following Aristotle and Thomas, he saw logic as the science on operating concepts concerning reality and not as a science on reality alone. Therefore, logic is the science *de entibus secundae intentionis*. However, this clearly opposes the objective concept of logic. Salamucha was aware of this difficulty but did not develop this issue.

These problems appeared because of the question concerning the applicability of mathematical logic to metaphysical issues. According to the scholastic tradition, mathematical logic is placed on the second level of abstraction whereas philosophy and in particular, metaphysics, on the third level. Salamucha did not reject this Medieval classification but sought a solution in the observation that Medieval mathematics and logic differed from modern mathematics and logic. In his paper „O możliwości ścisłego formalizowania dziedziny pojęć analogicznych" [On Possibilities of a Strict Formalization of the Domain of Analogical Notions] (1937a), he wrote that Medieval mathematics analysed the quantitative characteristics of objects whereas modern mathematics broke with this approach and "for the majority of modern mathematicians mathematics is simply a deductive theory, in which from some axioms and definitions derivative theorems are derived with the help of logical theses, mathematics can contain no empirical elements"[152] (2003, p. 79).

---

151 „Chociaż w ten sposób zapożyczam wiele od logików matematycznych, nie znaczy to wcale, że solidaryzuję się z ich nominalistycznym nastawieniem w logice i z materialistycznymi czy pozytywistycznymi tendencjami w filozofii. Myślę, że tak samo jak na gruncie logiki tradycyjnej mogły występować równie zgodnie, czy równie niezgodnie, różne systemy filozoficzne, podobnie sprawa się przedstawia na gruncie logiki matematycznej, tyle, że tu obowiązuje większa odpowiedzialność" (1934).
152 „dla większości współczesnych matematyków matematyka jest po prostu teorią dedukcyjną, w której z pewnych aksjomatów i definicji wyprowadza się przy pomocy tez logicznych pewne twierdzenia pochodne – żadnych elementów doświadczalnych matematyka zawierać nie może" (1937a, p. 132).

Thanks to that mathematics becomes similar to logic and along with the latter can – as noticed above – be treated as the science dealing with objects of second intention (cf. 1937a, p. 128). Salamucha adds (2003, p. 79):

> In this way, mathematics has got closer to logic to such an extent that the boundaries between what has till recently been two branches of sciences, today slowly disappear and mathematics becomes simply a part of logic, only higher and deductively later than those parts of the same science which are commonly regarded as logic.[153]

He summarizes his considerations (2003, p. 83):

> Thus, it appears that the fears that the application of logistic to metaphysics constitutes a violation of the differences between the traditional degrees of abstraction, are a result of some misunderstandings. Too great an emphasis has been laid upon the origin of logistic and modern mathematics has been confused with medieval mathematics.[154]

In Salamucha's opinion, logic is a theory of deductive argumentation. Unfortunately, he did not develop this idea. Therefore, it is not clear – as Koj writes (1995, p. 20) – "whether logic should be treated as a theory of consequences or whether only as metasystemic theses saying which objective theses should be accepted."[155] However, logic enables us to control reasoning. Reasoning as a mental activity is not intersubjectively verifiable, but through ordering expressions to particular elements of reasoning and through the analysis of the operations conducted on these expressions, we can check the conformity of inference with logical rules. Salamucha spoke here about methodological nominalism. It should be noted that it is something different than, e.g., Chwistek's nominalism,[156] which treats reasoning just as an operation on expressions (devoid of meaning). In Rev. Salamucha's opinion, logic does not exclude meanings but only temporarily – exactly for methodological reasons – abstracts from them while analysing the arguments. Yet, Salamucha stressed that such a conception of logic did not force nominalism in philosophical theories in which it is utilized.

---

153 „W ten sposób matematyka zbliżyła się do logiki do tego stopnia, że granice między tymi dwiema do niedawna gałęziami nauk dziś powoli się zacierają i matematyka staje się po prostu częścią logiki, wyższą tylko i dedukcyjnie późniejszą od tych części tej samej nauki, które powszechnie za logikę są uważane" (1937a, p. 132).

154 „Okazuje się, że obawy, jakoby zastosowanie logistyki do metafizyki było pogwałceniem różnic między tradycyjnymi stopniami abstrakcji, są wynikiem pewnych nieporozumień; kładzie się zbyt wielki nacisk na pochodzenie logistyki i miesza się matematykę współczesną z matematyką średniowieczną" (1937a, p. 137).

155 „czy logikę [należy] traktować jako teorię konsekwencji, czy tylko jako metasystemowe tezy mówiące, jakie tezy przedmiotowe należy przyjąć".

156 On Chwistek's philosophy of logic and mathematics and in particular on his nominalism see Murawski (2011a, 2011b).

One of the consequences of such a conception of logic is the thesis that logic is not creative but only consists in checking the conducted activities (for instance, reasonings); it allows checking and ordering deduction. At the same time, it is to some extent a universal science, i.e. its theses can be used in all disciplines. Salamucha wrote that "the normative consequences of logic embrace all fields of science and even ordinary life if we want it to be at least a little logical" (1936, p. 620).[157]

Salamucha did not claim that the formal logic of his times was a sufficient tool allowing the analysis and precise reconstruction of the whole of scholastic philosophy. When asked whether logic was such a sufficient tool, he said that he did not know that. Referring to *Principia Mathematica*, he claimed that it was sufficient to construct the whole of mathematics. However, he did not exclude the fact that in the future it would be necessary to enlarge logic so that we might use it to make adequate analyses of philosophical problems.[158] Salamucha realized that his investigations and those conducted by the Cracow Circle were something new and belonged to the domain which had not been developed before. Concluding his paper „O możliwości ścisłego formalizowania ..." he wrote (2003, p. 95):

> The arguments of this paper resemble forcing one's way through a jungle, where man rarely enters; logisticians who are not interested in scholasticism do not enter there–scholastics who are not interested in logistic do not enter there.[159]

In Salamucha's opinion, one of the problems that should be solved and developed was the issue related to analogy. The adversaries of the Cracow Circle raised the reservation – because of the use of formal logic in metaphysics – that the latter used analogous concepts whereas logic aimed at providing precise concepts and making them unambiguous. Rev. Salamucha did not have a solution for that but he saw that the concept of analogy, which scholastics used, was vague and pointed to some ideas of Drewnowski included in his work „Zarys programu filozoficznego" [Outline of a Philosophical Program] (1934). Consequently, he formulated the following interesting opinion (cf. 2003, p. 94):

---

157 „normatywne konsekwencje logiki obejmują wszystkie dziedziny naukowe i nawet życie potoczne, jeżeli chcemy, żeby ono było choć trochę logiczne".
158 This need was also presented clearly by Bocheński when he tried to formulate certain aspects of the problem of universals using the terms of modern logic – cf. Bocheński (1956b). He claimed that logical-mathematical investigations concerning certain questions connected with the problem of universals might require stronger logical and semantic tools than those that were available at that time.
159 „Wywody tego artykułu są trochę takie, jak przedzieranie się przez gąszcze, gdzie rzadko wdziera się człowiek; nie wchodzą tam logistycy, których scholastyka nie interesuje, – nie wchodzą tam scholastycy, którzy nie zajmują się logistyką" (1937a, p. 152).

It seems, however, that in metaphysics an adequately interpreted metalogic is going to be more useful than modern formal logic itself.[160]

As for the philosophical problems related to mathematics, we should also consider Salamucha's opinion that the appearance of non-Euclidean geometries and the creation of relativity theory allowed us to break down – as he wrote – the tyranny of time and space. He argued that both concepts were non-empirical and because of that we could not empirically confirm the influence of time and space on physical phenomena. Thanks to these new theories, the concept of space became "empirically reversed" and "space is only a conceptual construction and this construction can be undoubtedly and consistently developed in many different ways" (1946).[161] It is not clear what Salamucha meant by that. Since if geometry is to be understood as a formal science, experience does not play any role in recognizing its theorems as true or rejecting them as false. Both types of geometry – Euclidean and non-Euclidean – have the same epistemological status in this conception. However, if this or that geometry is used to construct physical models, experience will play a fundamental role here and will determine the adequacy of the description provided by the given model.

Finally, our reflection on Salamucha's views should include his praise of Roman Ingarden's criticism of the philosophy of the Vienna Circle and his answer to the question of using formal logic in phenomenology. Salamucha agrees with Ingarden to some extent, writing (2003, p. 84):

> If one claims, together with Prof. Ingarden, that all and only those issues belong to philosophy which concern either (a) "pure possibilities or necessary relations between possibilities" or (b) "the real existence of all possible domains of being" and "the real essence of both entire domains of being and their particular elements", where the main stress is laid upon the subject (a), then one will have to accept – at most with some small reservations – that the methods of particular sciences, and hence also the deductive method, will have no application to philosophy.[162]

---

160 „Zdaje się jednak, że w metafizyce bardziej przydatna okaże się metalogika, odpowiednio tylko interpretowana, aniżeli sama współczesna logika formalna" (1937a, p. 151).

161 „doświadczalnie wywracane"; „przestrzeń jest tylko konstrukcją pojęciową i można tę konstrukcję konsekwentnie i bezsprzecznie na różne sposoby rozbudowywać".

162 „Jeżeli się przyjmie, razem z prof. Ingardenem, że do filozofii należą te wszystkie zagadnienia i tylko te, które dotyczą: (a) czystych możliwości lub koniecznych związków między możliwościami lub (b) faktycznego istnienia wszelkich możliwych dziedzin bytu i faktycznej istoty zarówno całych dziedzin bytowych jak i ich poszczególnych elementów, przy tym główny nacisk położy się na tematach (a), to – co najwyżej z pewnymi małymi zastrzeżeniami – trzeba będzie uznać, że metody nauk szczegółowych, a więc i metoda dedukcyjna, nie znajdą w filozofii zastosowania" (1937a, p. 139).

However, this would lead – according to Salamucha – to a radical reduction of philosophical problems. Yet, if one wants to cultivate Thomistic philosophy and theology, the utilization of logistic tools is fully justified.

3. Let us proceed to the views concerning logic of another member of the Cracow Circle Jan F. Drewnowski. He formulated a more refined philosophical conception than the other members did – cf. his „Zarys programu filozoficznego" [Outline of a Philosophical Program] (1934), which became a kind of manifesto of the Cracow Circle although the other members of the Circle referred to it rather loosely. Drewnowski – as opposed to Salamucha or Bocheński – did not follow Thomism but chose his own way. In addition, he was an expert in natural sciences. His philosophical program was based on the interdependence of various fields of science, especially logic, natural sciences, mathematics and theology.

Drewnowski's aim was to propose a new philosophical language that could be used to express the views of many different philosophers, in particular the theses of modern scientific philosophical theories and the theses of classical philosophy, including Thomism.

One of the important components of Drewnowski's program was his theory of signs. In his opinion, signs play a substitutive role, allowing us to get to know the real world by going beyond direct sensations and by creating systems.[163] However, we should expect to face here certain threats, which Drewnowski specified in „Zarys" (1996, p. 58):[164]

> Falling into a habit of constant intercourse with signs instead of reality itself, meaning – so to say – an intentional attitude towards reality, causes in the long run that the sense of this intentionality is blurred.[165]

On the one hand, identifying signs with reality can reduce reality to what the signs define and on the other hand can recognize what the signs give as some new domain of reality.[166]

---

163 One can say that Drewnowski has anticipated in a way some of the ideas of Jacques Derrida.
164 All of the quotations come from „Zarys programu filozoficznego" included in Drewnowski's collection of selected works *Filozofia i precyzja* [Philosophy and Precision] (1996).
165 „Przyzwyczajenie do ciągłego obcowania ze znakami zamiast z samą rzeczywistością, czyli taki – że tak powiem – intencjonalny stosunek do rzeczywistości, sprawia na dalszą metę zatarcie się poczucia tej intencjonalności".
166 Twardowski also warned against this kind of errors (cf. his „Symbolomania i pragmatofobia" [Symbolomania and Pragmatophobia], 1927). In turn, Łukasiewicz recommended a constant contact with reality while using developed philosophical systems.

At first, signs substitute the reality under consideration and then help in theoretical considerations, which leads to the so-called pile-up of signs, i.e. groups of signs are replaced by other signs. Not paying attention to this problem may lead to a misunderstanding. Additionally, one should distinguish between signs and the instructions describing how to use these signs.

Drewnowski distinguished three kinds of theories: scientific, mathematical and theological. All of them are systems of signs. Drewnowski formulated rules of using signs for each kind of these theories and reflected on the relationships between the theories.

Having presented the general remarks, we can proceed to discussing Drewnowski's views on mathematics and logic as well as the applicability of logic to other sciences. Let us begin with his remarks on axioms and definitions. He claims that (1996, p. 67):

> Axioms are either expressions of certain presumptions of the so-called laws that are binding in a given domain or they are only expressions of certain agreements accepted within a given notation. In both cases they do not express anything absolute: in the first case – it is more correct to formulate them as suitable conditions and put them in an abbreviated way in the antecedents of the theorems of a theory;[167] in the other – they belong to regulatory instructions, and it is more correct to formulate them as appropriate directives.[168]

Drewnowski views definitions in a similar way.

In Drewnowski's approach mathematical theories are "the same sign mechanisms as other theories of natural sciences" (1996, p. 71).[169] He describes them more precisely in „Zarys" (1996, pp. 71–72):

> Their characteristics are that they are tools to analyse scientific theories themselves and all other systems of signs that look like scientific theories. They deal only with the properties of the construction of the systems of signs occurring in theories, namely the dependence of various structural types of complex signs on the ways of using them, in accordance with the regulatory instructions of a given theory. [...] Therefore, the only type of operations

---

167 We are dealing here with the theorem of deduction – my remark, R.M.
168 „Aksjomaty są wyrazem bądź pewnych przypuszczeń co do obowiązujących w danej dziedzinie tzw. praw, bądź´ też tylko są wyrazem pewnych umów przyjętych w obrębie danego znakowania. I w jednym, i w drugim wypadku nie wyrażają niczego bezwzględnego: w pierwszym – poprawniej jest sformułować je jako odpowiednie warunki i w skrócony sposób wymieniać je w poprzednikach twierdzeń teorii; w drugim wypadku – należą do instrukcji wykonawczej, i poprawniej jest sformułować je jako odpowiednie dyrektywy".
169 „są takimi samymi mechanizmami znakowymi, jak inne teorie przyrodnicze".

on mathematical theories is the operations that mark the deduction of propositions and similar inter-propositional relations.[170]

What is the relation of the commonly understood mathematics towards the mathematical theories thus characterized? Drewnowski claims that some parts of mathematics are scientific theories, in particular the arithmetic of natural numbers based on the primary concepts of quantity and sign. Natural theories include, in Drewnowski's opinion, "all geometries if they concern some extensive properties and do not move to generalizations, dealing with any relations of which a special case is a given relation occurring in some empirical extension" (1996, p. 73).[171] The remaining part of "contemporary mathematics can probably be comprised in the so-called theory of relations, i.e., it will depend on what I call here mathematical theories" (1996, p. 73).[172] In addition, for a mathematical theory it does not matter what the signs signify, and consequently, "the propositions of mathematics are devoid of any definite meaning" (*ibidem*).[173] The identification of mathematics with mathematical theories leads to the thesis that "objects with which mathematics deals are arbitrary human creations" (*ibidem*).[174] The problem of existence of mathematical objects can be reduced to the existence of signs which a given theory uses – as opposed to the scientific theories "where the indispensible condition of correctness, the verifiability of arguments, will always be to indicate the way of making available to the analysis what the theory is about" (1996, p. 74).[175]

According to Drewnowski, such mathematical theories include all the generalizations of philosophy and the whole part of metaphysics dealing with general laws. However, he regards the incompetent mathematization of various domains as wrong. At this point, it results from the schematization of mathematical domains "which do not know the dependencies that modern mathematics investigates" or

---

170 „Charakterystyczną cechą ich jest to, że są narzędziami do badania samych teorii przyrodniczych i wszelkich innych układów znaków, wyglądających jak teorie przyrodnicze. Zajmuja się one wyłącznie właściwościami budowy układów znaków występujących w teoriach, mianowicie tym, jak uzależnione są różne typy strukturalne znaków złożonych od sposobów posługiwania się nimi, zgodnie z instrukcjami wykonawczymi danej teorii. [...] Jedynym więc typem operacji na gruncie teorii matematycznych są te, które znaczą wywiedlność zdań i pokrewne zależności międzyzdaniowe".

171 „wszelkie geometrie o tyle, o ile zajmują się jakimiś własnościami rozciągłymi, a nie przechodzą do uogólnień zajmujących się dowolnymi stosunkami, których szczególnym przypadkiem bywa dany stosunek występujący w jakiejś rozciągłości doświadczalnej".

172 „współczesnej matematyki da się prawdopodobnie objąć tzw. teorią stosunków, czyli należeć będzie do tego, co nazywam tu teoriami matematycznymi".

173 „zdania matematyki są pozbawione określonego znaczenia".

174 „twory, którymi zajmuje się matematyka, są dowolnymi wytworami ludzkimi".

175 „gdzie zawsze niezbędnym warunkiem poprawności, sprawdzalności wywodów będzie wskazanie sposobu udostępniania badaniu tego, o czym mowa w teorii".

attempts "to transfer only mathematical symbols to various considerations, e.g., historical-philosophical ones, by those who do not know mathematics" (1996, p. 75).[176] The starting point of the correct mathematization of a theory must be suitable scientific theories based on empirical data – Drewnowski includes here "colour or tactile qualities serving as the starting point to construct the notions of physics, such as the sensations of pain, fear, adoration, the sense of ownership, of rightness, etc., which can serve as the starting points of many different scientific theories" (1996, p. 76).[177] Such theories can then be mathematized. Drewnowski describes this process as follows (1996, p. 76):

> It will consist in that: as the given scientific theory is being developed, its dependencies are getting more and more complicated; it will be stated that such dependencies are special relations, worked on in the mathematical theories. Then the whole part of a proper mathematical theory can be used in the given scientific theory by substituting the signs of the dependencies occurring in the scientific theory, which are special relations, analyzed in the mathematical theory, in the correct theorems of the mathematical theory. And reversely – various new dependencies in the given scientific theory can incline us to generalize them and thus provide new problems to the mathematical theories.[178]

Drewnowski regards the features of the mathematical theories as the advantages and benefits of this mathematization, writing (1996, p. 76):

> The value of this mathematization of knowledge will occur even more clearly when on the one hand, it is considered that the mathematical theories owe their efficiency to their higher degree of generality: analyzing the dependencies, without considering their meanings, allows making many attempts and modifications, which would not be easy within the

---

176 „którym obce są te zależności, jakimi zajmuje się współczesna matematyka"; "przenoszenia samych tylko symboli matematycznych do różnych rozważań, np. historiozoficznych, przez osoby nie znające matematyki".

177 „jakości barwne lub dotykowe, służące za punkt wyjścia do budowy pojęć fizyki, jak doznania bólu, strachu, uwielbienia, poczucia własności, słuszności itp., mogące służyć za punkty wyjścia szeregu innych teorii przyrodniczych".

178 „Będzie to polegać na tym, że w miarę rozwijania się danej teorii przyrodniczej, komplikacji występujących w niej zależności, stwierdzać się będzie, iż pewne takie zależności są szczególnymi przypadkami stosunków, opracowywanych w teoriach matematycznych. Wówczas cała ta część odpowiedniej teorii matematycznej może być zastosowana do danej teorii przyrodniczej drogą podstawienia w odpowiednich twierdzeniach teorii matematycznej znaków tych zależności teorii przyrodniczej, które są szczególnymi przypadkami stosunków badanych w teorii matematycznej. Odwrotnie też – różne nowe zależności w danej teorii przyrodniczej mogą skłaniać do uogólniania ich i dostarczać w ten sposób nowych zagadnień teoriom matematycznym".

framework of some scientific theory in which the meanings of signs, many a time loaded with tradition, habits, hinder the movements.[179]

On the other hand, considering the possible applications of the mathematical theories allows us to choose from "a surplus of possible combinations" those which are more desired.

Drewnowski also considered the problem of applying symbolic logic, especially to philosophy. He wrote a special paper on this question in 1965, referring to D. Hilbert and W. Ackermann's *Grundzüge der theoretischen Logik* (1928), which – as he notices – characterizes the method of this application of logic. He describes the method in the following words (1996, p. 199):[180]

> The method establishes constant symbols, expressing the specific notions of a given domain, and describes the types of objects marked by the arguments of these new functional symbols. With the help of these new symbols and the symbols of functional calculus,[181] the symbolic formulations of the premises from the given domain are provided. The formulated premises are added to the axioms of functional calculus as new axioms. From this, using the rules of inference of functional calculus, we receive theorems, being the symbolic formulations of what we want to prove in the given domain.[182]

At the same time, he notices that such an application of the predicate calculus is not an interpretation of the symbols of the language of this calculus because "all the time these symbols are used in the same general logical meaning as in the classical logical calculus" (1996, pp. 199–200).[183] The symbolic formulation of the assumed properties of the analysed objects in the form of axioms can feature certain general

---

179 „Wartość tak pojętego matematyzowania wiedzy wystąpi jeszcze wyraźniej, gdy się zważy, że z jednej strony teorie matematyczne zawdzięczają swoją sprawność większej swej ogólności: zajmowanie się zależnościami, bez oglądania się na ich znaczenie, pozwala na dokonywanie wielu prób i przeróbek, które nie byłyby łatwe w obrębie jakiejś teorii przyrodniczej, gdzie znaczenia znaków, obarczone nieraz tradycją, nawykami, utrudniają swobodę ruchów".

180 Like in the case of „Zarys programu filozoficznego", the page numbering is from Drewnowski's selected works *Filozofia i precyzja* (1996).

181 The old name of the predicate calculus – my remark, R.M.

182 „Metoda ta polega na tym, że ustala się nowe symbole stałe, wyrażające swoiste pojęcia danej dziedziny, i opisuje się rodzaje przedmiotów oznaczonych przez argumenty tych nowych symboli funkcyjnych. Za pomocą tych nowych symboli oraz symboli rachunku funkcyjnego podaje się symboliczne sformułowania przesłanek z danej dziedziny. Tak sformułowane przesłanki dołącza się do aksjomatów rachunku funkcyjnego jako nowe aksjomaty. Stąd zaś, stosując reguły wnioskowania rachunku funkcyjnego, otrzymuje się twierdzenia, będące symbolicznymi sformułowaniami tego, czego się chce dowieść w danej dziedzinie".

183 „symbole te cały czas są użyte w tym samym ogólnologicznym znaczeniu, jakie mają w klasycznym rachunku logicznym".

dependencies in a given domain, and the axioms "do not have to use up semantically the content of the notions and all the dependencies of this domain" (1996, p. 200).[184] Such an application of the logical tools to define precisely the given domain of knowledge does not violate the richness of its content. The application of these tools is possible to the extent that "the rational cognition of the given domain of reality" (1996, p. 200)[185] is possible. Moreover, Drewnowski clearly opposes the view that symbolic logic cannot be used outside of mathematics, in particular in philosophy. He criticizes the arguments formulated by the followers of this standpoint, especially the opinions presented by the adherents of the so-called existential Thomism who claim that "metaphysics cultivated in this spirit has separate methods of reasoning, and symbolic logic cannot be used here" (1996, pp. 200–201).[186] This problem was already considered by Ajdukiewicz in „O stosowalności czystej logiki do zagadnień filozoficznych" [On Applicability of Pure Logic to Philosophical Questions] (1934). He asked whether modern logic, which was extensional, could be used to solve philosophical problems formulated in the intentional colloquial language. In the aforementioned paper (1965), Drewnowski analysed three meanings of extensionality and stated that equivalential extensionality of classical logical calculus was not an obstacle to using this calculus in philosophy. He also explained (1996, pp. 203–204) how to use logic to solve philosophical and theological problems in the works of the Cracow Circle:

> All of our attempts neither interpreted logical symbols nor translated metaphysics into the language of symbolic logic. The method of applying symbolic logic, which we have utilized, was just [...] the application of the very classical logical calculus, to which new constant symbols are added.[187]

4. Let us proceed to the last member of the Cracow Circle – Fr Józef (Innocenty) Maria Bocheński. At this point, a certain problem is the evolution of his philosophical views since Bocheński was a follower of Kant, and then of neo-Thomism. He attempted to modernize the latter by using the tools of mathematical logic. Finally, he departed from the problem of being and moved towards analytic philosophy. Since our paper concerns the pre-war period, we are not going to analyse his post-war views but focus on his activities in the Cracow Circle (however, we will sometimes refer to his later activities).

---

184 „nie muszą wyczerpywać znaczeniowo treści pojęć i wszelkich zależności tej dziedziny".
185 „rozumne poznanie danej dziedziny rzeczywistości".
186 „metafizyka uprawiana w tym duchu ma odrębne metody rozumowania i logika symboliczna nie daje się tu stosować".
187 „Otóż wszystkie te nasze próby nie były ani interpretowaniem symboli logicznych, ani przekładaniem metafizyki na język logiki symbolicznej. Metoda stosowania logiki symbolicznej, jaką się posługiwaliśmy, była właśnie [...] stosowaniem samego tylko klasycznego rachunku logicznego, do którego dodaje się nowe symbole stałe".

According to the Cracow Circle, if Thomism wants to be a rational philosophy – which it has been since the very beginning – it must know and use modern formal logic. This logic gives precision, which Bocheński understood in the following way (1937, pp. 28–29):

> Our way of speaking is called "precise", when it observes the following rules: As far as words are concerned, they must be unequivocal signs of simple things, features, experiences, etc.; they are to be clearly defined in relation to these simple signs, in accordance with precisely stated rules. Furthermore, these words should be always used in such a way that each one of them constitutes a part of a proposition, i.e. expression that is true or false. Where propositions are concerned, they cannot be accepted until we know exactly what they mean and why we assent to them. Sometimes we accept them as evident, sometimes on the basis of faith or proof – in the latter case it should be conducted on the basis of clearly formulated and efficient logical directives.[188]

Additionally, precise speaking and thinking should be characterized by the use of formal logic and exclusion of such irrational factors as will, emotion and imagination. Bocheński was convinced that the best available logic is mathematical logic (formal logic, logistic) - cf., e.g. (1936) – but later he thought that certain philosophical problems required richer logical tools.

After the war, he rejected Thomism and followed analytic philosophy, being faithful to the discussed metaphilosophical principle and stressing the question of the method. Doubting whether one exact philosophical system, embracing all philosophical issues, could be built, he analysed unconnected problems separately, always using exact logical methods.

Refuting the accusations that were made during the discussion at the aforementioned meeting of the Cracow Circle in Kraków in August 1936, Fr Bocheński paid attention to the necessity of distinguishing between formal logic and philosophy as well as to the fact that in antiquity there had been logical systems different from Aristotle's logic. Similarly, in logistic it is the classical two-valued logic that plays a fundamental role. At the same time, formal logic does not focus so much on the truth of conclusions deduced by applying logical tools – it is the task of other

---

188 „Ścisłym nazywamy sposób mówienia, w którym obowiązują następujące zasady: Jeśli chodzi o użyte słowa, mają one być bądź niedwuznacznymi znakami prostych rzeczy, cech, doznań itp., bądź też być na gruncie poprawnie sformułowanych dyrektyw za pomocą takich właśnie znaków jasno zdefiniowane. Słowa te mają być dalej użyte zawsze tak, by każde z nich stanowiło część zdania, to jest wyrażenia, które jest prawdziwe albo fałszywe. Jeśli chodzi o zdania mogą one być uznane dopiero wtedy, gdyśmy sobie w pełni zdali sprawę, co znaczą i dlaczego je uznajemy. Racją tego uznania będzie niekiedy oczywistość, niekiedy wiara, niekiedy dowód – w ostatnim przypadku ma on być przeprowadzony na gruncie jasno sformułowanych i sprawnych dyrektyw logicznych".

sciences – but on the truth of its theses. He also stressed the possibility of using many-valued logics in theology. These logics could be treated as the logics of probability and utilized to evaluate the degrees of falsity – this may allow us to realize the idea of St Thomas Aquinas. Moreover, Bocheński claimed that the process of constructing logical systems did not assume any philosophical presumptions – logic is and should be neutral. The fact that mathematical logic grew out of mathematics, and like mathematics it uses symbolic notation, does not mean that formal logic can be used only in mathematics. It can and should be used wherever deduction is used – the deduction should be always exact and accurate.

When Bocheński cultivated philosophy in the spirit of analytic philosophy, he used the broadly understood logic, embracing formal logic as well as semiotics, which was based on it, and the general methodology of sciences. In his opinion, this conception of logic is an ideal pattern of rationality; it provides notional tools to analyse complex argumentations and to analyse notions. It constitutes the organon of philosophy and being a kind of ontology it constitutes a branch of philosophy.[189]

According to Bocheński modern logic is an autonomous science – but not only this kind of logic. In fact, in every epoch a highly developed logic had the right to be characterized as autonomous (cf. his paper of 1980). Asked whether modern logic is a mathematical discipline or whether it should be included in philosophy, he answers (1980) that it depends on the definitions of mathematics and philosophy. If mathematics is defined through its method, logic, using the same method and having the same characteristics (symbolic, formalistic, deductive, objective, etc.) as the mathematical sciences, should be regarded as a mathematical discipline. As a matter of fact, the boundaries between modern logic and mathematics are blurred. However, logic has two features that distinguish it from mathematics. The first one is the maximal generality of its fundamental branches such as propositional calculus, predicate calculus or the logic of relations; and the second is its higher degree of exactness (cf. Bocheński 1980, Section VI). On the other hand, assuming that philosophy analyses the foundations and most general properties of objects, modern logic, as any logic, becomes part of philosophy (*ibidem*, Section VII). This thesis is also supported – in Bocheński's opinion – by the fact that modern logic has given solutions to many traditional philosophical problems. He mentions Russell's conception of logical paradoxes and his theory of systematic ambiguity. This is – according to Bocheński – a solution of the eternal problem of the "univocity of being". Further, he mentions Tarski's definition of truth as well as Gödel's

---

189 Bocheński wrote in (1980): "In fact [modern logic] deals with "ultimate foundations" and it consists of axiomatic study of most abstract properties of any objects. Just therefore Heinrich Scholz claimed – rightly as it seems – that modern logic is an ontology being a fundamental part of philosophy".

first incompleteness theorem. He claims that the latter has shown among other things that there are no philosophical systems that could embrace the whole of reality (like Hegel's system). Thus logic, as a tool of philosophy, is also its part. This is also possible when logic is a part of mathematics, which results from – according to Bocheński – the fact that it is the most general and fundamental part of mathematics. It seems – he adds – that "the same truths are obligatory in the fundamental parts of all sciences" (Bocheński 1980, Section VII).

Note. The financial support of the National Center for Science [Narodowe Centrum Nauki] (Grant No N N101 136940) is acknowledged.

Bibliography

Achtner, W. (2005). Infinity in science and religion: the creative role of thinking about infinity, *Neue Zeitschrift für systematische Theologie und Geschichtsphilosophie* 47, 392–411.

Achtner, W. (2011). Infinity as a transformative concept in science and technology. In: M. Heller, W. Hugh Woodin (Eds.), *Infinity. New Research Frontiers*, Cambridge University Press, Cambridge, pp. 19–51.

Ackermann, W. (1924–1925). Begründung des tertium non datur mittels der Hilbertschen Theorie der Widerspruchsfreiheit, *Mathematiche Annalen* 93, 1–36.

Ackermann, W. (1940). Zur Widerspruchsfreiheit der Zahlentheorie, *Mathematische Annalen* 117, 162–194.

Aigner, M., Ziegler, G.M. (1998). *Proofs From the Book*, Springer Verlag, Berlin–Heidelberg–New York. Second edition: 2001, third edition: 2004.

Ajdukiewicz, K. (1934). O stosowalności czystej logiki do zagadnień filozoficznych, *Przegląd Filozoficzny* 37, 323–327.

Albertson, D. (2014). *Mathematical Theologies: Nicholas of Cusa and the Legacy of Thierry of Chartres*, Oxford University Press, New York.

Apt, K.R., Marek, W. (1974). Second order arithmetic and related topics, *Annals of Mathematical Logic* 6, 177–239.

Aschbacher, M. (2005). Highly complex proofs and implications of such proofs, *Philosophical Transactions of the Royal Society* A, 363, 2401–2406.

Avigad, J. (2006). Mathematical method and proof, *Synthese* 153, 105–159.

Banach, S., Tarski A. (1924). Sur la décomposition des ensembles de points en parties respectivement congruentes, *Fundamenta Mathematicae* 6, 244–277.

Barwise, J. (1977). An introduction to first-order logic. In: J. Barwise J. (Ed.), *Handbook of Mathematical Logic*, North-Holland, Amsterdam New York/Oxford, pp. 5–46.

Bassler, G.B. (2006). The surveyability of mathematical proof: a historical perspective, *Synthese* 148, 99–133.

Batóg, T. (1996). *Dwa paradygmaty matematyki. Studium z dziejów i filozofii matematyki*, Wydawnictwo Naukowe Uniwersytetu im. Adama Mickiewicza, Poznań.

Becker, O. (1927). Mathematische Existenz. Untersuchungen zur Logik und Ontologie mathematischer Phänomene, *Jahrbuch für Philosophie und phänemenologische Forschung* 8, 439–809.

Bedürftig, Th., Murawski, R. (2015). *Philosophie der Mathematik*, Walter de Gruyter, Berlin/Boston (3rd edition); Berlin/Boston 2019 (4th edition).

Bedürftig, Th., Murawski, R. (2018), *Philosophy of Mathematics*, Walter de Gruyter, Berlin/Boston.
Bendiek, J. (1949). Scholastische und mathematische Logik, *Franziskanische Studien* 31, 13–48.
Bendiek, J. (1956). Zur logischen Struktur der Gottesbeweise, *Franziskanische Studien* 38, 1–38, 296–321.
Bernays, P. (1967). Hilbert David. In: P. Edwards (Ed.), *Encyclopedia of Philosophy*, vol.3, Macmillan/Free Press, New York, pp. 496–504.
Betti, A. (2008). Polish axiomatics and its truth. On Tarski's Lesniewskian background and the Ajdukiewicz connection. In: D. Patterson (Ed.), *New Essays on Tarski and Philosophy*, Oxford University Press, Oxford, pp. 44–71.
Bocheński, J.M. (1936). W sprawie logistyki, *Verbum* 3, 445–454.
Bocheński, J.M. (1937). Tradycja myśli katolickiej a ścisłość. In: *Myśl katolicka wobec logiki współczesnej, Studia Gnesnensia* 15, 27–34.
Bocheński, J.M. (1938). Analisi logica di un testo di S. Tomasso d'Aquino (I,75,6). In: *Nove lezioni dilogica simbolica*, Angelium, Roma, pp. 147–155.
Bocheński, J.M. (1948). On analogy, *The Thomist* 11, 474–497. Reprinted in: A. Menne (Ed.), *Logico-philosophical studies*, D. Reidel, Dordrecht 1962, pp. 96–117.
Bocheński, J.M. (1956a). *Formale Logik*, Verlag Karl Alber, Freiburg–München.
Bocheński, J.M. (1956b). The problem of universals. In: J.M. Bocheński, A. Church, N. Goodman (Eds.), *The Problem of Universals*. University of Notre Dame Press, Notre Dame, pp. 33–54.
Bocheński, J.M. (1980). The general sense and character of modern logic. In: E. Agazzi (Ed.), *Modern logic – a Survey*. D. Reidel, Dordrecht, pp. 3–14.
Bocheński, J.M. (1989). The Cracow Circle. In: K. Szaniawski (Ed.), *The Vienna Circle and the Lvov – Warsaw School*. Kluwer Academic Publishers, Dordrecht, pp. 9–18.
Bornstein, B. (1912). *Prolegomena filozoficzne do geometryi*, Skład Główny w Księgarni E. Wende i S-ka (T. Hiż and A. Turkuł), Warszawa.
Bornstein, B. (1913). Problemat istnienia linji geometrycznych, *Przegląd Filozoficzny* 16, 64–73.
Bornstein, B. (1914). Podstawy filozoficzne teorji mnogości, *Przegląd Filozoficzny* 17, 183–193.
Bornstein, B. (1915). W sprawie recenzji p. Stanisława Leśniewskiego rozprawy mojej pt. „Podstawy filozoficzne teorji mnogości", *Przegląd Filozoficzny* 18, 121–140.
Bornstein, B. (1916). *Elementy filozofii jako nauki ścisłej*, Skład Główny w Księgarni E. Wendego i S-ka, Warszawa.

Bornstein, B. (1922). Zarys architektoniki i geometrji świata logicznego, *Przegląd Filozoficzny* 25, 475–490.
Bornstein, B. (1926). Geometrja logiki kategorialnej i jej znaczenie dla filozofii, *Przegląd Filozoficzny* 29, 173–194.
Bornstein, B. (1928). *La logique géométrique et sa porte philosophique*, Bibliotheca Universitatis Liberae Polonae, Warszawa.
Bornstein B. (1935). *Architektonika świata*, vol. II: *Logika geometryczno-architektoniczna*, Skład Główny Gebethner i Wolff, Warszawa.
Bornstein, B. (1939). *Geometrical Logic: the Structures of Thought and Space*, Wolna Wszechnica, Warszawa.
Bornstein, B. (1946). *Zarys teorii logiki dialektycznej*. Unpublished, hand-written copy in the Jagiellonian Library, Kraków.
Bornstein, B. (a). *O kategorialnej geometrii algebraicznej*. Unpublished, hand-written copy in the Jagiellonian Library, Kraków.
Bornstein, B. (b). *Co to jest kategorialna geometria algebraiczno-logiczna*. Unpublished, hand-written copy in the Jagiellonian Library, Kraków.
Bornstein, B. (c). *Geometria analityczna Descartes'a a geometria filozoficzna*. Unpublished, hand-written copy in the Jagiellonian Library, Kraków.
Bourbaki, N. (1968). *Theory of Sets*, Addison-Wesley, Reading, MA.
Breidert, W. (1977). Mathematik und symbolische Erkenntnis bei Nikolaus von Kues, *Mitteilungen und Forschungsbeiträge der Cusanus-Gesellschaft* 12, 116–126.
Browder, F.E. (Ed.) (1976). *Mathematical Developments Arising from Hilbert's Problems*, Proceedings of Symposia in Pure Mathematics vol. XXVIII, Part 1 and 2, American Mathematical Society, Providence, RI.
Brown, D.K. (1987). *Functional analysis in weak subsystems of second-order arithmetic*, Ph.D. Thesis, Pennsylvania State University, University Park, PA.
Brown, D.K., Simpson, S.G. (1986). Which set existence axioms are needed to prove the separable Hahn–Banach theorem?, *Annals of Pure and Applied Logic* 31, 123–144.
CadwalladerOlsker, T. (2011). What do we mean by mathematical proof?, *Journal of Humanistic Mathematics* 1, 33–60.
Cantor, G. (1895). Beiträge zur Begründung der transfiniten Mengenlehre, *Mathematische Annalen* 46, 481–512.
Carnap, R. (1952). *The Logical Foundations of Probability*, University of Chicago Press, Chicago, IL.
Carnap, R. (1963). Intellectual autobiography. In: P.A. Schilpp (Ed.), *The Philosophy of Rudolf Carnap*, Open Court, La Salle, IL, pp. 3–84.
Church, A. (1936). An unsolvable problem of elementary number theory, *American Journal of Mathematics* 58, 345–363. Reprinted in: Davis (1965), pp. 89–107.

Chwistek, L. (1917). Trzy odczyty odnoszące się do pojęcia istnienia, *Przegląd Filozoficzny* 20, 122–151. Reprinted in Chwistek (1961), pp. 3–29.

Chwistek, L. (1921). *Wielość rzeczywistości*, Ministerstwo Wyznań Religijnych i Oświecenia Publicznego, Kraków. Reprinted in: Chwistek (1961), 30–105.

Chwistek, L. (1923). Zastosowanie metody konstrukcyjnej do teorii poznania, *Przegląd Filozoficzny* 26, 175–187.

Chwistek, L. (1924). The theory of constructive types (principles of logic and mathematics). Part I: General principles of logic: theory of classes and relations, *Annales de la Société Polonaise de Mathématique* II, 9–48.

Chwistek, L. (1925). The theory of constructive types (principles of logic and mathematics). Part II: Cardinal arithmetic, *Annales de la Société Polonaise de Mathématique* III, 92–141.

Chwistek, L. (1933.) *Zagadnienia kultury duchowej w Polsce*, Gebethner and Wolff, Warszawa. Reprinted in Chwistek (1961), pp. 149–277.

Chwistek, L. (1935). *Granice nauki. Zarys logiki i metodologii nauk ścisłych*, (Książnica Atlas, Lwów/Warszawa. Reprinted in Chwistek (1963), pp. 1–232. English translation: Chwistek (1948).

Chwistek, L. (1948). *The Limits of Science. Outline of Logic and of the Methodology of the Exact Sciences*. English translation by H. C. Brodie, and A. P. Coleman, Kegan Paul, Trench Trubner, New York/London. Reprinted by Routledge, London 2000.

Chwistek, L. (1961). *Pisma filozoficzne i logiczne*, vol. I, Państwowe Wydawnictwo Naukowe, Warszawa.

Chwistek, L. (1963). *Pisma filozoficzne i logiczne*, vol. II, Państwowe Wydawnictwo Naukowe, Warszawa.

Clark, J.T. (1952). *Conventional logic and modern logic. A prelude to transition*, Woodstock College Press, Woodstock.

Counet, J.-M. (2005). Mathematics and the divine in Nicholas of Cusa. In: T. Koetsier and L. Bergmans (Eds.), *Mathematics and the Divine: a Historical Study*, Elsevier, Amsterdam, pp. 273–290.

Couturat, L. (1905). *L'Algebre de la logique*, Gauthier-Villars, Paris. English translation: *The Algebra of Logic*, The Open Court Publishing Company, Chicago 1914.

Czeżowski, T. (1918). Imiona i zdania. Dwa odczyty, *Przegląd Filozoficzny* 21, 3–22.

Davis, M. (Ed.) (1965). *The Undecidable: Basic Papers on Undecidable Propositions, Unsolvable Problems, and Computable Functions*, Raven Press, Hewlett, NY.

DeLong, H. (1970). *A Profile of Mathematical Logic*, Addison-Wesley, Reading, MA.

Detlefsen, M. (1979). On interpreting Gödel's second theorem, *Journal of Philosophical Logic* 8, 297–313.

Detlefsen, M. (1990). On the alleged refutation of Hilbert's program using Gödel's first incompleteness theorem, *Journal of Philosophical Logic* 19, 343–377.

Detlefsen M. (1993). The Kantian character of Hilbert's formalism: Kant's ideas and Hilbert's ideals. In: J. Czermak (Ed.), *Philosophy of Mathematics. Proceedings of the 15th International Wittgenstein-Symposium*, Part I, Verlag Hölder-Pilcher-Tempsky, Wien, pp. 195–205.

De Villiers, M.D. (1999). *Rethinking Proof with the Geometer's Sketchpad*, Key Curriculum Press, Emeryville, CA.

Drake, F.F. (1989). On the foundations of mathematics in 1987. In: H.-D. Ebbinghaus et al. (Eds.), *Logic Colloquium'87*, Elsevier Science Publishers B.V., Amsterdam, pp. 11–25.

Drewnowski, J.F. (1934). Zarys program filozoficznego, *Przegląd Filozoficzny* 37, 3–38, 150–181 and 262–292. Reprinted in: Drewnowski (1996), pp. 55–147.

Drewnowski, J.F. (1965). Stosowanie logiki symbolicznej w filozofii, *Studia Philosophiae Christianae* 1, 53–65.

Drewnowski, J.F. (1996). *Filozofia i precyzja. Zarys programu filozoficznego i inne pisma*, Wydawnictwo Towarzystwa Naukowego Katolickiego Uniwersytetu Lubelskiego, Lublin.

Federici Vescovini, G. (1997). Cusano e la matematica. In: *Filosofia e storia della cultura. Studi in onore di Fulvio Tessitore*, vol. 1: *Dall'antico al moderno*, Morano, Napoli, pp. 392–434.

Federici Vescovini, G. (1998). L'architettonica della mente di Cusano e la matematica. In: *Le origini della modernità*, a cura di W. Tega, vol. I: *Linguaggi e saperi tra XV e XVI secolo*, Firenze, pp. 31–48.

Feferman, A.B., Feferman, S. (2004). *Alfred Tarski. Life and Logic*. Cambridge University Press, Cambridge.

Feferman, S. (1964–1968). Subsystems of predicative analysis, Part I: *Journal of Symbolic Logic* 29 (1964), 1–30; part II: *Journal of Symbolic Logic* 33 (1968), 193–220.

Feferman, S. (1984). Kurt Gödel: conviction and causation. In: *Philosophia Naturalis*, a special issue, P. Weingartner et al. (Eds.) *Philosophy of Science – History of Science. A Selection of Contributed Papers of the 7th International Congress of Logic, Methodology and Philosophy of Science, Salzburg, 1983*, Verlag Anton Hain, Meisenheim/Glan. Reprinted in: S.G. Shanker (Ed.) *Gödel's Theorem in Focus*, Croom Helm, London, 1988, pp. 96–114.

Feferman, S. (1988). Hilbert's program revisited: proof–theoretical and foundational reductions, *Journal of Symbolic Logic* 53, 364–384.

Field, H. (1994). Deflationist views of meaning and content, *Mind* 103, 249–285.

Folina, J. (2006). Church's Thesis and the variety of mathematical justifications. In: A. Olszewski *et al.* (Eds.), *Church's Thesis After 70 Years*, Ontos Verlag, Frankfurt am Main, pp. 220–241.
Føllesdal, D. (1995). Gödel and Husserl. In: J. Hintikka (Ed.), *Essays on the Development of the Foundations of Mathematics*, Kluwer Academic Publishers, Dordrecht, pp. 427–446.
Frege G. (1976). *Wissenschaftlicher Briefwechsel*, Hrsg. G. Gabriel, H. Hermes, F. Kambartel, Ch. Thiel, A. Veraart, Felix Meiner Verlag, Hamburg.
Friedman, H. (1975). Some subsystems of second order arithmetic and their use. In: *Proceedings of the International Congress of Mathematicians*, Canadian Mathematical Congress, Vancouver, vol. 1, pp. 235–242.
Friedman, H. (1976). Systems of second order arithmetic with restricted induction (abstracts), *Journal of Symbolic Logic* 41, 557–559.
Friedman, H. (1981). On the necessary use of abstract set theory, *Advances in Mathematics* 41, 209–280.
Friedman, H., Simpson, S.G., Smith, R.L. (1983). Countable algebra and set existence axioms, *Annals of Pure and Applied Logic* 25, 141–181. Addendum: *Annals of Pure and Applied Logic* 28 (1985), 319–320.
Frost-Arnold, G. (2004). Was Tarski's theory of truth motivated by physicalism?, *History and Philosophy of Logic* 25, 265–280.
Gandy, R. (1988). Confluence of ideas in 1936. In: R. Herken (Ed.), *The Universal Turing Machine – a Half-Century Survey*, Oxford University Press, New York, pp. 55–111.
Gödel, K. (1929). Über die Vollständigkeit des Logikkalküls, doctoral dissertation, submitted in 1929; published and translated in: Gödel (1986), pp. 60–101.
Gödel, K. (1930). Die Vollständigkeit der Axiome des logischen Funktionenkalküls, *Monatshefte für Mathematik und Physik* 37, 349–360.
Gödel, K. (1931). Über formal unentscheidbare Sätze der *Principia Mathematica* und verwandter Systeme. I, *Monatshefte für Mathematik und Physik* 38, 173–198. Reprinted with English translation: On formally undecidable propositions of *Principia Mathematica* and related systems, in: Gödel (1986), pp. 144–195.
Gödel, K. (1931?). Über unentscheidbare Sätze; first published (German text and English translation 'On undecidable sentences') in: Gödel (1995), pp. 30–35.
Gödel, K. (1933). The present situation in the foundations of mathematics; first published in: Gödel (1995), pp. 45–53.
Gödel, K. (1934). *On Undecidable Propositions of Formal Mathematical Systems* (mimeographed lecture notes, taken by S.C. Kleene and J.B. Rosser), Princeton; reprinted with revisions in: Davis (1965), pp. 39–74.

Gödel K. (1944). Russell's mathematical logic. In: P.A. Schilpp (Ed.), *The Philosophy of Bertrand Russell*, Northwestern University, Evanston, pp. 123–153. Reprinted in: Gödel (1990), pp. 119–141.

Gödel, K. (1946). Remarks before the Princeton Bicentennial Conference on problems in mathematics; first published in: Davis (1965), pp. 84–88; reprinted in Gödel (1990), pp. 150–153.

Gödel, K. (1947/1964). What is Cantor's continuum problem?, *The American Mathematical Monthly* 54, 515–525. Second revised version in: P. Benacerraf and H. Putnam (Eds.), *Philosophy of Mathematics. Selected Readings*, Prentice-Hall, Englewood Cliffs, NJ, 1964, pp. 258–273.

Gödel, K. (1951). Some basic theorems on the foundations of mathematics and their implications; first published in: Gödel (1995), pp. 304–323.

Gödel, K. (1953). Is mathematics syntax of language? (unfinished contribution); first published in: Gödel (1995), pp. 334–362.

Gödel, K. (1958). Über eine bisher noch nicht benützte Erweiterung des finiten Standpunktes, *Dialectica* 12, 280–287.

Gödel, K. (1961). The modern development of the foundations of mathematics in the light of philosophy; first published (German text and English translation) in: Gödel (1995), pp. 374–387.

Gödel, K. (1972). On an extension of finitary mathematics which has not yet been used; revised and expanded English version of (Gödel 1958), to have appeared in *Dialectica*, first published in: Gödel (1990), pp. 271–280.

Gödel, K. (1986). *Collected Works*, S. Feferman *et al.* (Eds.), vol. I, Oxford University Press/Clarendon Press, Oxford/New York.

Gödel, K. (1990). *Collected Works*, S. Feferman *et al.* (Eds.), vol. II, Oxford University Press/Clarendon Press, Oxford/New York.

Gödel, K. (1995). *Collected Works*, S. Feferman *et al.* (Eds.), vol. III, Oxford University Press/Clarendon Press, Oxford/New York.

Grattan-Guinness, I. (2000). *The Search for Mathematical Roots 1870–1940. Logics, Set Theories and the Foundations of Mathematics from Cantor through Russell to Gödel*, Princeton University Press, Princeton/London.

Grover, D. (1992). *A Prosentential Theory of Truth*, Princeton University Press, Princeton, NJ.

Hartimo, M. (2017). Husserl and Gödel's incompleteness theorems, *Review of Symbolic Logic* 10(4), 638–650.

Heijenoort, J. van (Ed.) (1967). *From Frege to Gödel: A Source Book in Mathematical Logic, 1879–1930*, Harvard University Press, Cambridge, MA.

Hersh, R. (1997). *What Is Mathematics, Really?*, Oxford University Press, New York.

Hilbert D. (1899). *Grundlagen der Geometrie. Festschrift zur Feier der Enthüllung des Gauss-Weber-Denkmals*, B.G. Teubner, Leipzig, pp. 3–92.

Hilbert D. (1900). Über den Zahlbegriff, *Jahresbericht der Deutschen Mathematikervereinigung* 8, 180–184.
Hilbert, D. (1901). Mathematische Probleme, *Archiv der Mathematik und Physik* 1, 44–63 and 213–237. Reprinted in: Hilbert (1935), pp. 290–329. English translation: Mathematical problems, *Bulletin of the American Mathematical Society* 8 (1901–1902), 437–479. Also in: F. Browder (Ed.), *Mathematical Developments Arising from Hilbert's Problems*, Proceedings of the Symposia in Pure Mathematics 28, American Mathematical Society, Providence, RI, 1976, pp. 1–34.
Hilbert D. (1902/1903). Über den Satz von der Gleichheit der Basiswinkel im gleichschenkligen Dreieck, *Proceedings of the London Mathematical Society* 35, 50–68.
Hilbert D. (1903). *Grundlagen der Geometrie*, 2nd edition, Teubner, Leipzig.
Hilbert D. (1905a). Logische Principien des mathematischen Denkens, Lecture notes by Ernst Hellinger, Mathematisches Institut, Georg-August-Universität Göttingen, Sommer-Semester 1905. Unpublished manuscript.
Hilbert D. (1905b). Über die Grundlagen der Logik und der Arithmetik. In: A. Krazer (Ed.), *Verhandlungen des dritten Internationalen Mathematiker-Kongresses in Heidelberg vom 8. bis 13. August 1904*, Teubner, Leipzig, pp. 174–185.
Hilbert D. (1917–1918). Prinzipien der Mathematik, Lecture notes by Paul Bernays. Mathematisches Institut, Georg-August-Universität Göttingen, Wintersemester 1917–18. Unpublished typescript.
Hilbert D. (1918). Axiomatisches Denken, *Mathematische Annalen* 78, 405–415.
Hilbert, D. (1926). Über das Unendliche, *Mathematische Annalen* 95, 161–190; English translation: On the infinite, in: J. van Heijenoort (Ed.) (1967), pp. 367–392.
Hilbert, D. (1927). Die Grundlagen der Mathematik, *Abhandlungen aus dem mathematischen Seminar der Hamburgischen Universität* 6, 65–85. Reprinted in: D. Hilbert, *Grundlagen der Geometrie*, 7th edition, Leipzig, 1930. English translation: The foundations of mathematics, in: J.van Heijenoort (Ed.) (1967), pp. 464–479.
Hilbert D. (1930a). Probleme der Grundlegung der Mathematik, *Mathematische Annalen* 102, 1–9.
Hilbert, D. (1930b). Naturerkennen und Logik, *Naturwissenschaften* 18, 959–963.
Hilbert, D. (1931). Die Grundlegung der elementaren Zahlentheorie, *Mathematische Annalen* 104, 485–494.
Hilbert D. (1935). *Gesammelte Abhandlungen*, Band 3, Springer-Verlag, Berlin.
Hilbert, D., Ackermann, W. (1928). *Grundzüge der theoretischen Logik*, Julius Springer, Berlin.
Hilbert, D., Bernays, P. (1934–1939). *Grundlagen der Mathematik*, Springer Verlag, Berlin, Band I: 1934, Band II: 1939.

Hiż, H. (1966). Kotarbiński on truth. In: *Studies in Polish Civilization*, D. Wandycz (Ed.), Columbia University Press, New York, pp. 426–431.
Horwich, P. (1999). *Truth*, 2nd edition, Oxford University Press, New York.
Husserl, E. (1891). *Philosophie der Arithmetik. Psychologische und logische Untersuchungen*, C.E.M. Pfeffer, Halle-Saale.
Husserl, E. (1900–1901), *Logische Untersuchungen*, vols. 1–2, Niemeyer, Halle.
Husserl, E. (1970). *Philosophie der Arithmetik. Mit Ergänzenden Texten (1890–1901)*, Martinus Nijhoff, The Hague.
Husserl, E. (1994). *Edmund Husserl Briefwechsel*. Band VII: *Wissenschaftskorrespondenz*, Kluwer Academic Publishers, Dordrecht.
Husserl, E. (2003). *Philosophy of Arithmetic. Psychological and Logical Investigations with Supplementary Texts from 1887–1901. Collected Works*, vol. X, Kluwer Academic Publishers, Dordrecht.
Isaacson, D. (1987). Arithmetical truth and hidden higher-order concepts. In: The Paris Logic Group (Eds.), *Logic Colloquium'85*, Elsevier Science Publishers B.V. (North-Holland), Amsterdam, pp. 147–169.
Isaacson, D. (1992). Some considerations on arithmetical truth and the ω-rule. In: M. Detlefsen (Ed.) *Proof, Logic and Formalization*, Routledge, London/New York, pp. 94–138.
Janiszewski, Z. (1917). O potrzebach matematyki w Polsce, *Nauka polska, jej potrzeby, organizacja i rozwój* 1, 11–18. Reprinted in: *Roczniki Polskiego Towarzystwa Matematycznego*, Seria II: *Wiadomości Matematyczne* 7, 3–8. English translation in: Kuzawa (1968).
Kahle, R. (2015). What is a proof?, *Axiomathes* 25, 79–91.
Kalmár, L. (1959). An argument against the plausibility of Church's Thesis. In: A. Heyting (Ed.), *Constructivity in Mathematics, Proceedings of the Colloquium Held at Amsterdam 1957*, North-Holland, Amsterdam, pp. 72–80.
Kauferstein, C. (2006). *Transzendentalphilosophie der Mathematik*, ibidem-Verlag, Stuttgart.
Kaufmann, F. (1930). *Das Unendliche in der Mathematik und seine Ausschaltung*, Franz Deuticke, Leipzig/Vienna.
Kaye, R. (1991). *Models of Peano Arithmetic*, Clarendon Press, Oxford.
Kirby, L., Paris, J. (1977). Initial segments of models of Peano's axioms. In: A. Lachlan, M. Srebrny, A. Zarach (Eds.), *Set Theory and Hierarchy Theory V. Bierutowice, Poland, 1976*, Lecture Notes in Mathematics 619, Springer Verlag, Berlin-Heidelberg-New York, pp. 211–226.
Kirby, L., Paris, J. (1982). Accessible independence results for Peano arithmetic, *Bulletin of the London Mathematical Society* 14, 285–293.
Kleene, S.C. (1952). *Introduction to Metamathematics*, Noordhoff, Groningen.
Kleene, S.C. (1967). *Mathematical Logic*, John Wiley & Sons, New York.

Kleene, S.C. (1987). Reflections on Church's thesis, *Notre Dame Journal of Formal Logic* 28, 490–498.

Knobloch, E. (2002). Unendlichkeit und Mathematik bei Nicolaus aus Kues – Grundideen und ihre Weiterentwicklung. In: A. Schürmann und B. Weiss (Hrsg.), *Chemie – Kultur – Geschichte. Festschrift für Hans-Werner Schütt anlässlich seines 65. Geburtstages*, Verlag für Geschichte der Naturwissenschaften und der Technik, Berlin-Diepholz, pp. 223–234.

Koj, L. (1995). Ks. Jana Salamuchy koncepcja logiki. In: Z. Wolak (red.) *Logika i metafizyka*, Biblos, Tarnów – OBI, Kraków, pp. 15–31.

Kokoszyńska, M. (1936). Über den Absoluten Wahrheitsbegriff und einige andere semantische Begriffe, *Erkenntnis* 6, 143–165. Reprinted in: *Logische Rationalismus: Philosophische Schriften der Lemberg-Warschauer Schule*, D. Pearce and J. Woleński (Eds.), Athenäum, Frankfurt/Main 1988, pp. 276–292.

Kokoszyńska, M. (1948). What means a "relativity" of truth, *Studia Philosophica* 3, 167–175.

Kokoszyńska, M. (1951. A refutation of the relativism of truth, *Studia Philosophica* 4, 93–149.

Kotarbińska, J. (1984). Głos w dyskusji, *Studia Filozoficzne* 5 (222), 69–73.

Kotarbiński, T. (1926). *Elementy logiki formalnej, teorji poznania i metodologji*, authorized manuscript, D. Steinberżanka (Ed.), Wydawnictwo Koła Filozoficznego S.U.W. i Koła Przyrodników S.U.W, Warszawa.

Kotarbiński, T. (1929). *Elementy teorji poznania, logiki formalnej i metodologji nauk*, Ossolineum, Lwów. 2nd enlarged edition: Ossolineum, Wrocław–Warszawa–Kraków 1961. English translation (from 2nd edition): Kotarbiński (1966).

Kotarbiński, T. (1934). W sprawie pojęcia prawdy, *Przegląd Filozoficzny* 37, 85–91.

Kotarbiński, T. (1935). Zasadnicze myśli pansomatyzmu, *Przegląd Filozoficzny* 38, 283–294. English translation by A. Tarski and D. Rynin: Kotarbiński (1955).

Kotarbiński, T. (1955). The fundamental ideas of pansomatism, *Mind* 64, 488–500, *Mind* 65, 288. Reprinted in: Tarski (1986), vol. 3, pp. 577–591.

Kotarbiński, T. (1966). *Gnosiology: the Scientific Approach to the Theory of Knowledge*, Pergamon, Oxford.

Kotlarski, H. (1986). Bounded induction and satisfaction classes, *Zeitschrift für Mathematische Logik und Grundlagen der Mathematik* 32, 531–544.

Kotlarski H., Ratajczyk, Z. (1990a). Inductive full satisfaction classes, *Annals of Pure and Applied Logic* 47, 199–223.

Kotlarski H., Ratajczyk, Z. (1990b). More on induction in the language with a full satisfaction class, *Zeitschrift für Mathematische Logik und Grundlagen der Mathematik* 36, 441–454.

Krajewski, S. (1976). Non-standard satisfaction classes. In: W. Marek, M. Srebrny, A. Zarach (Eds.), *Set Theory and Hierarchy Theory*, Proceedings of the

Bierutowice Conference 1975, Lecture Notes in Mathematics 537, Springer Verlag, Berlin/Heidelberg/New York, pp. 121–144.

Krajewski, S. (2006). Remarks on Church's theses and Gödel's theorem. In: A. Olszewski *et al.* (Eds.), *Church's Thesis After 70 Years*, Ontos Verlag, Frankfurt am Main, pp. 269–280.

Kreisel, G. (1958). Hilbert's programme, *Dialectica* 12, 346–372; reprinted with Postscript in: P. Benacerraf and H. Putnam (Eds.), *Philosophy of Mathematics*, Prentice–Hall, Englewood Cliffs, NJ, 1964, pp. 157–180.

Kreisel, G. (1970). Church's thesis: a kind of reducibility axiom for constructive mathematics. In: J. Myhill, A. Kino, and R.E. Vesley (Eds.), *Intuitionism and Proof Theory*, North-Holland, Amsterdam, pp. 121–150.

Kuratowski, K., Mostowski, A. (1952). *Teoria mnogości*, Nakładem Polskiego Towarzystwa Matematycznego z subwencji Ministerstwa Szkolnictwa Wyższego, Warszawa/Wrocław. Second edition: Kuratowski, Mostowski (1966), Third edition: Kuratowski, Mostowski (1978). English translation: Kuratowski, Mostowski (1967).

Kuratowski, K., Mostowski A. (1966). *Teoria mnogości*, 2nd edition, Państwowe Wydawnictwo Naukowe, Warszawa.

Kuratowski, K., Mostowski, A. (1967). *Set Theory*, PWN (Polish Scientific Publishers) and North-Holand, Warszawa/Amsterdam. Second edition 1976.

Kuratowski, K., Mostowski, A. (1978). *Teoria mnogości*, 3rd edition, Państwowe Wydawnictwo Naukowe, Warszawa.

Kuzawa, S.M.G. (1968). *Modern Mathematics: The Genesis of a School in Poland*, College & University Press, New Haven.

Leśniewski, S. (1914). Teoria mnogości na „podstawach filozoficznych" Benedykta Bornsteina, *Przegląd Filozoficzny* 17, 488-507. Reprinted: O podstawach filozoficznych teorii mnogości, *Filozofia Nauki* 2(22) (1998), 123–139.

Leśniewski, S. (1929). Grundzüge eines neuen Systems der Grundlagen der Mathematik, *Fundamenta Mathematicae* 14, 1–81. English translation: Fundamentals of a New System of the Foundations of Mathematics, in: S. Leśniewski *Collected Works*, S.J. Surma, J. Strzednicki and D.I. Barnett (Eds.), Kluwer Academic Publishers, Dordrecht, pp. 410–605.

Lusin, N. (1983). List do Arnolda Denjoy z 1926 r., *Roczniki Polskiego Towarzystwa Matematycznego*, Seria II: *Wiadomości Matematyczne* 25, 65–68.

Łukasiewicz, J. (1910). *O zasadzie sprzeczności u Arystotelesa*, Polska Akademia Umiejętności, Kraków. German translation *Über den Satz des Widerspruchs bei Aristoteles*, Georg Olms Verlag, Hildesheim.

Łukasiewicz, J. (1911). Zagadnienie prawdy, *Księga Pamiątkowa XI Zjazdu Lekarzy i Przyrodników Polskich, Nakładem Towarzystwa Gospodarczego*, Kraków, 84–86. Reprinted in: Łukasiewicz (1998), 55–56.

Łukasiewicz, J. (1913). Die logischen Grundlagen der Wahrscheinlichkeitsrechnung, Polska Akademia Umiejętności; partial English translation: Logical foundations of probability theory, in: Łukasiewicz (1970), 16–63.

Łukasiewicz, J. (1915). O nauce. In: *Poradnik dla samouków*, vol. I, Heflich i Michalski, Warszawa, pp. 15–34. Reprinted in: Łukasiewicz (1998), 9–33. Partial English translation: On creative elements in science, in: Łukasiewicz (1970), 1–15.

Łukasiewicz, J. (1916). O pojęciu wielkości (Z powodu dzieła Stanisława Zaremby), *Przegląd Filozoficzny* 19, 1–70. English translation: On the concept of magnitude. In connection with Stanisław Zaremba's work, in: Łukasiewicz (1970), pp. 16–83.

Łukasiewicz, J. (1927). O metodę w filozofii, *Przegląd Filozoficzny* 31, 3–5.

Łukasiewicz, J. (1936). Logistyka a filozofia, *Przegląd Filozoficzny* 39, 115–131. English translation: Logistic and philosophy, in: Łukasiewicz (1970), 218–235.

Łukasiewicz, J. (1951). *Aristotle's Syllogistic from the Standpoint of Modern Formal Logic*, Clarendon Press, Oxford. Second edition (including the chapter on syllogistic of modal sentences) – Clarendon Press, Oxford 1957. Polish translation: *Sylogistyka Arystotelesa z punktu widzenia współczesnej logiki formalnej*, Państwowe Wydawnictwo Naukowe, Warszawa 1988.

Łukasiewicz, J. (1970). *Selected Works*, L. Borkowski (Ed.), North-Holland/Panstwowe Wydawnictwo Naukowe, Amsterdam/Warszawa.

Łukasiewicz, J. (1998). *Logika i metafizyka. Miscellanea*, J.J. Jadacki (Ed.), Wydział Filozofii i Socjologii Uniwersytetu Warszawskiego, Warszawa.

Maddy, P. (1980). Perception and mathematical intuition, *The Philosophical Review* 89(2), 163–196.

Maligranda, L. (2009). Eustachy Żyliński (1889–1954), *Antiquitates Mathematicae* 3, 171–211.

Mancosu, P. (2000). On mathematical explanation. In: E. Grosholz and H. Breger (Eds.), *The Growth of Mathematical Knowledge*, Kluwer Academic Publishers, Dordrecht, pp. 103–119.

Mancosu, P. (2001). Mathematical explanation: problems and prospects, *Topoi* 20, 97–117.

Mancosu, P. (2005). Harvard 1940–1941: Tarski, Carnap and Quine on a finitistic language of mathematics for science, *History and Philosophy of Logic* 26, 327–357.

Mancosu, P., Ryckman Th. (2005). Geometry, physics and phenomenology: four letters of O. Becker to H. Weyl. In: V. Peckhaus (Ed.), *Oskar Becker und die Philosophie der Mathematik*, Fink Verlag, München, pp. 229–243.

Marciszewski, W., Murawski, R. (1995). *Mechanization of Reasoning in a Historical Perspective*, Editions Rodopi, Amsterdam/Atlanta, GA.

Marczewski, E. (1948). Rozwój matematyki w Polsce, Nakładem Polskiej Akademii Umiejętności z zasiłku Prezydium Rady Ministrów, Skład Główny w Księgarni Gebethnera i Wolffa, Warszawa/Kraków/Łódz/Poznan/Zakopane.
Mazur, S. (1963). Computable Analysis, *Rozprawy Matematyczne* 33, 111 pp. [edited by A. Grzegorczyk and H. Rasiowa].
McCarty, C. (1987). Intuitionism and computability, *Notre Dame Journal of Formal Logic* 28, 536–580.
Mendelson, E. (1970). *Introduction to Mathematical Logic*, D. Van Nostrand, Princeton, NJ.
Mendelson, E. (1990). Second thoughts about Church's Thesis and mathematical proofs, *The Journal of Philosophy* 87, 225–233.
Mendelson, E. (2006). On the impossibility of proving the "hard-half" of Church's Thesis. In: A. Olszewski *et al.* (Eds.), *Church's Thesis After 70 Years*, Ontos Verlag, Frankfurt am Main, pp. 304–309.
Mostowski, A. (1955a). The present state of investigations of the foundations of mathematics, *Rozprawy Matematyczne* 9, 1–48, co-authors: A. Grzegorczyk, S. Jaśkowski, J. Łoś, S. Mazur, H. Rasiowa and R. Sikorski.
Mostowski, A. (1955b). Współczesny stan badań nad podstawami matematyki, *Prace Matematyczne* 1, 13–55.
Mostowski, A. (1959). On various degrees of constructivism. In: A. Heyting (Ed.), *Constructivity in Mathematics. Proceedings of the Colloquium held in Amsterdam, 1957*, North-Holland, Amsterdam, pp. 178–194.
Mostowski, A. (1965). *Thirty Years of Foundational Studies. Lectures on the Development of Mathematical Logic and the Study of the Foundations of Mathematics in 1930–1964*, volume fasc. XVII of *Acta Philosophica Fennica*. Reprinted as: Mostowski (1966) and in Mostowski (1979), pp. 1–176.
Mostowski, A. (1966). *Thirty Years of Foundational Studies. Lectures on the Development of Mathematical Logic and the Study of the Foundations of Mathematics in 1930–1964*, Basil Blackwell, Oxford.
Mostowski, A. (1967). Tarski Alfred. In: P. Edwards (Ed.), *The Encyclopedia of Philosophy*, vol. 8, Macmillan, New York, pp. 77–81.
Mostowski, A. (1972). Matematyka a logika. Refleksje przy lekturze książki A. Grzegorczyka *Zarys arytmetyki teoretycznej* wraz z próbą recenzji, *Roczniki Polskiego Towarzystwa Matematycznego*, Seria II: *Wiadomości Matematyczne* 15, 79–89.
Mostowski, A. (1975). Travaux de W. Sierpiński sur la théorie des en sembles et ses applications. In: W. Sierpiński (Ed.), *OEuvres choisies*, vol. 2, Państwowe Wydawnictwo Naukowe – Éditions Scientifiques de Pologne, Warszawa, pp. 9–13.
Mostowski, A. (1979). *Foundational Studies. Selected Works*, vols. I-II, Państwowe Wydawnictwo Naukowe/North Holland, Warszawa/Amsterdam.

Murawski, R. (1976–1977). On expandability of models of Peano arithmetic, I–III, *Studia Logica* 35 (1976), 409–419; 35 (1976), 421–431; 36 (1977), 181–188.

Murawski, R. (1984a). Expandability of models of arithmetic. In: G. Wechsung (Ed.), *Proceedings of Frege Conference 1984*, Akademie-Verlag, Berlin, pp. 87–93.

Murawski, R. (1984b). Matematyczna niezupełność arytmetyki, *Roczniki Polskiego Towarzystwa Matematycznego*. Seria II: *Wiadomości Matematyczne* 26, 47–58.

Murawski, R. (1997). Satisfaction classes – a survey. In: R. Murawski and J. Pogonowski (Eds.), *Euphony and Logos*, Editions Rodopi, Amsterdam–Atlanta, GA, pp. 259–281.

Murawski R. (1998). Undefinability of truth. The problem of the priority: Tarski vs. Gödel, *History and Philosophy of Logic* 19, 153–160.

Murawski, R. (1999). *Recursive Functions and Metamathematics. Problems of Completeness and Decidability, Gödel's Theorems*, Kluwer Academic Publishers, Dordrecht.

Murawski, R. (2002a). Truth vs. provability – philosophical and historical remarks, *Logic and Logical Philosophy* 10, 93–117.

Murawski, R. (2002b). On the distinction proof-truth in mathematics. In: P. Gärdenfors et al. (Eds.), *In the Scope of Logic, Methodology and Philosophy of Science*, Kluwer Academic, Dordrecht/Boston/London, pp. 287–303.

Murawski, R. (2004a). Philosophical reflection on mathematics in Poland in the interwar period, *Annals of Pure and Applied Logic* 127, 325–337.

Murawski, R. (2004b). Church's thesis and its epistemological status, *Annales Universitatis Mariae Curie-Skłodowska*, Sectio AI, Informatica, 2, 57–70.

Murawski, R. (2010). Philosophy of mathematics in the Warsaw Mathematical School, *Axiomathes* 20, 279–293.

Murawski, R. (2011a). *Filozofia matematyki i logiki w Polsce międzywojennej*, Monografie Fundacji na rzecz Nauki Polskiej, Wydawnictwo Naukowe Uniwersytetu Mikołaja Kopernika, Toruń.

Murawski, R. (2011b). On Chwistek's philosophy of mathematics. In: N. Griffin, B. Linsky and K. Blackwell, (Eds.). *Principia mathematica at 100*, The Bertrand Russell Centre, Hamilton, ON, pp. 121–130. Also in: *Russell: The Journal of Bertrand Russell Studies* 31 (2011), 121–130.

Murawski, R. (2014). *Philosophy of Logic and Mathematics in Poland in the 1920s and 1930s*, Birkhäuser Verlag, Basel.

Murawski, R., Świrydowicz, K. (2006). *Podstawy logiki i teorii mnogości*, Wydawnictwo Naukowe Uniwersytetu im. Adama Mickiewicza, Poznań. Second edition: Poznań, 2016.

Murawski, R., Woleński, J. (2006). The status of Church's thesis. In: Woleński J., Olszewski A. and Janusz R. (Eds.), *Church's Thesis After 70 Years*, Ontos, Frankfurt, pp. 310–330.

Murawski, R., Woleński, J. (2008). Andrzej Mostowski on the foundations and philosophy of mathematics. In: A. Ehrenfeucht, V.M. Marek and M. Srebrny (Eds.), *Andrzej Mostowski and Foundational Studies*, IOS Press, Amsterdam/Berlin/Oxford/Tokyo/Washington, DC, pp. 324–337.

Mycielski, J. (2004). On the tension between Tarski's nominalism and his model theory (Definitions for a mathematical model of knowledge), *Annals of Pure and Applied Logic* 126, 215–224.

Nicolai de Cusa. (1440). *De docta ignorantia*. In: *Opera omnia*, vol. I, ediderunt: E. Hoffmann, R. Klibansky, Lipsiae 1932. Novam editionem curavit Burkhard Mojsisch 2008. English translation: *On Learned Ignorance*, The Arthur J. Banning Press, Minneapolis, Minnesota 1985 (second edition), 1990 (second printing). Polish translation: *O oświeconej niewiedzy*, Wydawnictwo Znak, Kraków 1997.

Nicolai de Cusa. (1450). *Idiota de mente*. In: *Opera omnia*, vol. V, edidit R. Steiger, Hamburgi 1983. Novam editionem curavit Burkhard Mojsisch 2008. English translation: *Layman on Mind*, in: *Nicholas of Cusa On Wisdom and Knowledge*, The Arthur J. Banning Press, Minneapolis, Minnesota, 1996. Polish translation (with Latin text): *Laik o umyśle*, Wydawnictwo Marek Derewecki, Kęty 2008.

Nicolai de Cusa. (1458). *De mathematica perfectione*. In: Editio Argentoratensis 1488, vol. II, 490–498. Nova edition: P. Wilpert, Berlin 1967.

Nicolai de Cusa. (1463) *De venatione sapientiae. De apice theoriae*. In: *Opera omnia*, vol. XII, ediderunt: R. Klibansky, H.-G. Singer, Hamburg 1982. English translation: *On the Pursuit of Wisdom*. In: *Nicholas of Cusa: Metaphysical Speculations*, The Arthur J. Banning Press, Minneapolis, Minnesota, 1998.

Niebergall, K.-G. (1996). *Zur Metamathematik nichtaxiomatisierbarer Theorien*, Centrum für Informations- und Sprachverarbeitung Ludwig-Maximilians-Universität München, Bericht 96–87.

Nieznański, E. (1987). Logical analysis of Thomism. The Polish programme that originated in 1930's. In: J. Srzednicki (Ed.), *Initiatives in logic*, Martinus Nijhoff Publishers, Dordrecht/Boston/Lancaster, pp. 128–155. Reprinted in: Salamucha (2003), pp. 363–393.

Paris, J., Harrington L. (1977). A mathematical incompleteness in Peano arithmetic. In: J. Barwise (Ed.), *Handbook of Mathematical Logic*, North-Holland, Amsterdam, pp. 1133–1142.

Parsons, Ch. (1980). Mathematical intuition, *Proceedings of the Aristotelian Society* 80, 145–168.

Pasquerella, L. (1989). Brentano and Kotarbiński on truth. In: *The Object and Its Identity*, Kluwer Academic Publishers, Dordrecht, pp. 98–106.

Post, E. (1965). Absolutely unsolvable problems and relatively undecidable propositions – account of an anticipation. In: M. Davis (Ed.) (1965), pp. 340–433.

Reprinted in: E. Post, *Solvability, Provability, Definability: Collected Works*, M. Davis (Ed.), Birkhäuser, Basel 1994, pp. 375–441.

Prawitz, D. (1983). *Philosophical aspects of proof theory*. In: G. Fløstad (Ed.), *Contemporary Philosophy. a New Survey*, Martinus Nijhoff Publishers, The Hague/Boston/London, pp. 235–277.

Quine, W.V.O. (2004). *Philosophy of Logic*, 2nd edition, Harvard University Press, Cambridge, MA.

Ramsey, F. P. (1927). Facts and propositions, *Aristotelian Society Supplementary Volume* 7, 153–170.

Resnik, M.D. (1974). On the philosophical significance of consistency proofs, *Journal of Philosophical Logic* 3, 133–147.

Rogers, H. (1987). *Theory of Recursive Functions and Effective Computability*, McGraw-Hill, New York.

Rojszczak, A. (2005). *From the Act of Judging to the Sentence*, Kluwer Academic Publishers, Dordrecht.

Rota, G.-C. (1997). The phenomenology of mathematical proof, *Synthese* 111, 183–196.

Rowe, D.E. (1989). Klein, Hilbert, and the Göttingen mathematical tradition, *Osiris* 2 (5), 186–213.

Ryle, G. (1953). Ordinary language, *The Philosophical Review* XLII, 252–271; reprinted in: G. Ryle, *Collected Essays 1929-1968*, vol. 2, Thoemmes, Bristol, pp. 301–319.

Salamucha, J. (1934). Dowód ex motu na istnienie Boga. Analiza logiczna argumentacji św. Tomasza z Akwinu, *Collectanea Theologica* 15, 53–92. English translation: The proof ex motu for the existence of God. Logical analysis of St. Thomas' arguments, *The New Scholasticism* 32 (1958), 327–372. New English translation: The proof ex motu for the existence of God. Logical analysis of St. Thomas Aquinas' arguments, in: Salamucha (2003), pp. 97–135.

Salamucha, J. (1935). Logika zdań u Wilhelma Ockhama, *Przegląd Filozoficzny* 38, 208–239; German translation: Die Aussagenlogik bei Wilhelm Ockham, *Franziskanische Studien* 32 (1950), 97–134. English translation: The propositional logic in William Ockham, in: Salamucha (2003), pp. 139–167.

Salamucha, J. (1936). Zza kulis filozofii chrześcijańskiej, *Verbum* 3, 613–627. Reprinted in: Salamucha (1997), pp. 181–186.

Salamucha, J. (1937a). O możliwości ścisłego formalizowania dziedziny pojęć analogicznych, *Myśl katolicka wobec logiki współczesnej*, Studia Gnesnensia 15, 122–153. English translation: On possibilities of a strict formalization of the domain of analogical notions, in: Salamucha (2003), pp. 71–95.

Salamucha, J. (1937b). Pojawienie się zagadnień antynomialnych na gruncie logiki średniowiecznej, *Przegląd Filozoficzny* 40, 68–89, 320–343. English translation:

The appearance of antinomial problems with medieval logic, in: Salamucha (2003), pp. 169–210.
Salamucha, J. (1946). Czas, przestrzeń i wieczność, Dziś i Jutro 2 17(23), 3–4. Reprinted in: Salamucha (1997), pp. 83–88.
Salamucha, J. (2003). *Knowledge and Faith*, K. Świętorzecka and J.J. Jadacki, (Eds.), Rodopi, Amsterdam/New York.
Scholz, H. (1937). Besprechung: Studia Philosophica I, *Deutsche Literaturzeitung* 58, 1914–1917.
Schulz, K.-D. (1997). *Die These von Church. Zur erkenntnistheoretischen und sprachphilosophischen Bedeutung der Recursionstheorie*, Peter Lang, Frankfurt am Main.
Schütte, K. (1977). *Proof Theory*, Springer Verlag, Berlin/Heidelberg/New York.
Shapiro, S. (1983). Understanding Church's thesis, *Journal of Philosophical Logic* 10, 353–365.
Shapiro, S. (1993). Understanding Church's thesis, again, *Acta Analytica* 11, 59–77.
Shoenfield, J.R. (1967). *Mathematical Logic*, Addison-Wesley, Reading, MA.
Sierpiński, W. (1912). *Zarys teorii mnogości*, Skład Główny w Księgarni E. Wendego i S-ki, Warszawa.
Sierpiński, W. (1918). L'axiome de M. Zermelo et son role dans la théorie des ensembles et l'analyse, *Bulletin International de l'Académie des Sciences et des Lettres de Cracovie, classe de Sciences mathéematiques et naturelles*, Série A: *sciences mathématiques*, 97–152. Reprinted in: Sierpiński (1975), pp. 208–255.
Sierpiński, W. (1923). *Zarys teorii mnogości*, Wydawnictwo Kasy im. J. Mianowskiego, Warszawa.
Sierpiński, W. (1965). *Cardinal and Ordinal Numbers*, Polish Scientific Publishers, Warszawa.
Sierpiński, W. (1975). *Oeuvres choisies*, vol. 2, Państwowe Wydawnictwo Naukowe, Warszawa.
Simpson, S.G. (1984). Which set existence axioms are needed to prove the Cauchy/Peano theorem for ordinary differential equations?, *Journal of Symbolic Logic* 49, 783–802.
Simpson, S.G. (1985). Friedman's research on subsystems of second order arithmetic. In: L. Harrington et al. (Eds.), *Harvey Friedman's Research in the Foundations of Mathematics*, North-Holland, Amsterdam, pp. 137–159.
Simpson, S.G. (1988). Ordinal numbers and the Hilbert's basis theorem, *Journal of Symbolic Logic* 53, 961–974.
Simpson, S.G. (1998). *Subsystems of Second Order Arithmetic*, Springer-Verlag, New York. Second edition: Cambridge University Press, Cambridge, 2010.
Sinaceur, H.B. (2009). Tarski's practice and philosophy: between formalism and pragmatism. In: S. Lindström, K. Segerberg and V. Stoltenberg-Hansen (Eds.),

Logicism, Intuitionism, and Formalism: What Has Become of Them?, Springer Verlag, Berlin, pp. 355–394.

Skolimowski, H. (1967). *Polish Analytical Philosophy*, Roudedge and Kegan Paul, London.

Sleszyński, J. (1925–1929). *Teorja dowodu* [Wykłady uniwersyteckie oprac. S. K. Zaremba], vols. I–II, Kółko Matematyczno-Fizyczne Uczniów UJ, Kraków.

Smoryński, C. (1977). The incompleteness theorems. In: J. Barwise (Ed.), *Handbook of Mathematical Logic*, North-Holland, Amsterdam, pp. 821–865.

Smoryński, C. (1985). *Self-Reference and Modal Logic*, Springer Verlag, New York/Berlin/Heidelberg/Tokyo.

Smoryński, C. (1988). Hilbert's programme, *CWI Quarterly* 1, 3–59.

Smullyan, R. (1993). *Recursion Theory for Metamathematicians*, Oxford University Press, Oxford.

Suszko, R. (1968). Review of Andrzej Mostowski's *Thirty years of foundational studies. Lectures Notes on the Development of Mathematical Logic and the Study of the Foundations of Mathematics in 1930–1964*, Helsinki 1965, *Studia Logica* 22, 169–170.

Steiner, M. (1978). Mathematical explanation, *Philosophical Studies* 34, 133–151.

Steinhaus, H. (1923). *Czem jest a czem nie jest matematyka*, Księgarnia Nakładowa H. Altenberga, Lwów.

Steinhaus, H. (2000). *Między duchem a materią pśredniczy matematyka*, Wydawnictwo Naukowe PWN, Warszawa/Wrocław.

Suppes, P. (1988). Philosophical implications of Tarski's work, *Journal of Symbolic Logic* 53, 80–91.

Sylvan, R. (1997). *Transcendental Metaphysics*, The White Horse Press, Cambridge.

Śleziński, K. (2009). *Benedykta Borsteina koncepcja naukowej metafizyki i jej znaczenie dla badań współczesnych*, Wydawnictwo „scriptum", Kraków.

Śleziński, K. (2010). Bornsteinowska koncepcja podstaw teorii mnogości. Krytyka Stanisława Leśniewskiego i polemika Benedykta Bornsteina, *Studia z Filozofii Polskiej* 5, 97–112.

Tait, W.W. (1981). Finitism, *Journal of Philosophy* 78, 524–546.

Takeuti, G. (1987). *Proof Theory*, North-Holland, Amsterdam.

Tarski, A. (1930). Fundamentale Begriffe der Methodologie der Deduktiven Wissenschaften, *Monatshefte für Mathematik und Physik* 37, 361–404. English translation: Fundamental concepts of the methodology of the deductive sciences, in: Tarski (1956), pp. 60–109.

Tarski, A. (1932). Der Wahrheitsbegriff in den Sprachen der deduktiven Disziplinen. *Akademie der Wissenschaften in Wien, mathematisch-naturwissenschaftliche Klasse, Akademische Anzeiger* 69, 23–25. Reprinted in: Tarski (1986), pp. 613–617.

Tarski, A. (1933). Pojęcie prawdy w językach nauk dedukcyjnych, Towarzystwo Naukowe Warszawskie, Warszawa, Wydział III Nauk Matematyczno-Fizycznych, vol. 34. Reprinted in: A. Tarski, *Pisma logiczno-filozoficzne*, vol. 1: *Prawda*, Wydawnictwo Naukowe PWN, Warszawa 1995, pp. 131–172. English translation: *The concept of truth in formalized languages*. In: Tarski (1956), pp. 152–278 and in Tarski (1983), pp. 152–283.

Tarski, A. (1944). The semantic conception of truth and the foundation of semantics, *Philosophy and Phenomenological Research* 4, 341–376. Reprinted in: Tarski (1986), vol. 2, pp. 665–669.

Tarski, A. (1953). A general method in proofs of undecidability'. In: A. Tarski, A. Mostowski and R.M. Robinson (Eds.), *Undecidable Theories*, North-Holland, Amsterdam, pp. 1–35.

Tarski, A. (1954). Contribution to the discussion of P. Bernays 'Zur Beurteilung der Situation in der beweistheoretischen Forschung', *Revue Internationale de Philosophie* 8, 16–20.

Tarski, A. (1956). *Logic, Semantics, Metamathematics. Papers from 1923 to 1938*, Clarendon Press, Oxford. Second edition: Tarski (1983).

Tarski, A. (1969). Truth and proof, *Scientific American* 220(6), 63–77. Reprinted in: Tarski (1986), vol. 4, pp. 399–423.

Tarski, A. (1983). *Logic, Semantics, Metamathematics. Papers from 1923 to 1938*, 2nd edition edited and introduced by J. Corcoran, Hackett Publishing, Indianapolis.

Tarski, A. (1986) *Collected Papers*, vols. 1–4, S.R. Givant and R.N. McKenzie (Eds.), Birkhäuser, Basel.

Tarski, A. (1992). Drei Briefe an Otto Neurath |25.IV.1930, 10.VI.1936, 7.IX.1936|, edited by R. Haller, translated by J. Tarski, *Grazer Philosophische Studien* 43, 1–31.

Tarski, A. (2000.) Address at the Princeton University Bicentennial Conference on Problems of Mathematics (December 17–19, 1946), H. Sinaceur (Ed.), *The Bulletin of Symbolic Logic* 1, 1–44.

Tieszen, R. (1988). Phenomenology and mathematical knowledge, *Synthese* 75(3), 373–403.

Tieszen, R. (1994). The philosophy of arithmetic: Frege and Husserl. In: L. Haaparanta (Ed.), *Mind, Meaning and Mathematics*, Kluwer Academic Publishers, Dordrecht, pp. 85–112.

Twardowski, K. (1894). *Zur Lehre vom Inhalt und Gegenstand der Vorstellungen. Eine psychologische Untersuchung*, Alfred Hölder, Wien. Second edition: Philosophia Verlag, München/Wien, 1982. English translation: *On the Content and Object of Presentations*, Nijhoff, The Hague, 1976.

Twardowski, K. (1900). O tzw. prawdach względnych. In: *Księga Pamiątkowa Uniwersytetu Lwowskiego ku uczczeniu pięćsetnej rocznicy Fundacji Jagiellońskiej*,

Uniwersytet Lwowski, Lwów, pp. 64-93. English translation: On so-called elative truths, in: Twardowski (1999), pp. 147-169.

Twardowski, K. (1912). O czynnościach i wytworach. Kilka uwag z pogranicza psychologii, gramatyki i logiki. In: *Księga Pamiątkowa ku uczczeniu 250 rocznicy załozenia Uniwersytetu Lwowskiego przez króla Jana Kazimierza*, Uniwersytet Lwowski, Lwów, pp. 1-33; English translation: Actions and products: some remarks from the boderline of psychology, grammar and logic, in: Twardowski (1999), pp. 103-132.

Twardowski, K. (1927). Symbolomania i pragmatofobia. In: *Rozprawy i artykuły filozoficzne*, Lwów, pp. 394-406. Reprinted in: *Wybrane pisma filozoficzne*, Państwowe Wydawnictwo Naukowe, Warszawa 1965, pp. 354-363.

Twardowski, K. (1975). Teoria poznania (wykłady akademickie w r. a. 1924/25), *Archiwum Historii Filozofii i Myśli Społecznej* 21, 241-299. English translation: Theory of knowledge: a lecture course 1924/25, in: Twardowski (1999), pp. 103-132.

Twardowski, K. (1997). *Dzienniki*, tom 1: 1915-1927, tom 2: 1928-1936, do druku przygotował, wprowadzeniem i przypisami opatrzył R. Jadczak, Wydawnictwo Adam Marszałek, Warszawa-Toruń.

Twardowski, K. (1999). *On Actions, Products and Other Topics in Philosophy*, J. Brandl and J. Woleński (Eds.), Rodopi, Amsterdam.

Tymoczko, T. (1979). The four-color problem and its philosophical significance, *The Journal of Philosophy* 76, 57-83.

Vuissoz, F. (1998). *La conception sémantique de la vente. Logique et philosophie chez Alfred Tarski*, Centre de Recherches Semiologiques, Université de Neuchâtel, Neuchâtel.

Wang Hao (1974). *From Mathematics to Philosophy*, Routledge and Kegan Paul, London.

Wang Hao (1987). *Reflections on Kurt Gödel*, M.I.T. Press, Cambridge, MA.

Wang, Hao (1996). *A Logical Journey: From Gödel to Philosophy*, MIT Press, London, England/Cambridge, MA.

Webb, J. (1980). *Mechanism, Mentalism, and Metamathematics*, D. Reidel, Dordrecht.

Weyl, H. (1918). *Das Kontinuum. Kritische Untersuchungen über die Grundlagen der Analysis*, Veit, Leipzig.

Weyl, H. (1922). *Raum, Zeit, Materie. Vorlesungen über allgemeine Relativitätstheorie*, 5th edition, Springer, Wien.

Weyl, H. (1967). Comments on Hilbert's second lecture on the foundations of mathematics. In: J. van Heijenoort (Ed.) (1967), pp. 480-484.

Whitehead, A. N., Russell, B. (1925-1927). *Principia Mathematica*, 2nd edition, vols. I-III, The University Press, Cambridge.

Wiedijk, F. (2008). Formal proof – getting started, *Notices of the American Mathematical Society* 55, 1408–1414.
Wilkosz, W. (1938). *Liczę i myślę. Jak powstała liczba*, Księgarnia Powszechna, Kraków. Second edition: Państwowe Zakłady Wydawnictw Szkolnych, Warszawa, 1951.
Wolak, Z. (1993). *Neotomizm a Szkoła Lwowsko-Warszawska*, Ośrodek Badań Interdyscyplinarnych, Kraków.
Wolak, Z. (1996). Zarys historii Koła Krakowskiego. In: Z. Wolak (red.), *Logika i metafizyka*, Biblos, Tarnów – OBI, Kraków, pp. 79–84.
Wolak, Z. (2005). *Koncepcje analogii w Kole Krakowskim*, Wydawnictwo, Diecezji Tarnowskiej Biblos, Tarnów.
Woleński, J. (1985). *Filozoficzna Szkoła Lwowsko-Warszawska*, Państwowe Wydawnictwo Naukowe, Warszawa.
Woleński, J. (1989). *Logic and Philosophy in the Lvov-Warsaw School*, Kluwer Academic Publishers, Dordrecht/Boston/London.
Woleński, J. (1990). Kotarbiński, many-valued logic and truth. In: J. Woleński (Ed.), *Kotarbiński: Logic, Semantics and Ontology*, Kluwer Academic Publishers, Dordrecht, pp. 190–197. Reprinted in: Woleński (1999), 115–120.
Woleński, J. (1991). Gödel, Tarski and the undefinability of truth. In: *Yearbook 1991 of the Kurt Gödel Society*, Wien, pp. 97–108.
Woleński, J. (1992). Filozofia logiki i matematyki w warszawskiej szkole logicznej. In: *Matematyka przełomu XIX i XX wieku. Nurt mnogościowy*, Uniwersytet Śląski, Katowice, pp. 16–25.
Woleński, J. (1993). Tarski as a philosopher. In: F. Coniglione, R. Poli, J. Woleński (Eds.), *Polish Scientific Philosophy: The Lvov-Warsaw School*, Editions Rodopi, Amsterdam/New York, pp. 319–338.
Woleński, J. (1993b). Two concepts of correspondence, *From the Logical Point of View* 2, 42–55.
Woleński, J. (1994a). Jan Łukasieweicz on the Liar Paradox, logical consequence, truth and induction, *Modem Logic* 4, 392–400. Reprinted in: Woleński (1999), 121–125.
Woleński, J. (1994b). Theories of truth in Austrian philosophy, <http://www.fmag.unict.it/ polphil/Polphil/LvovWarsaw/WolTruth.html>. Reprinted in: Woleński (1999), 150–175.
Woleński, J. (1995). On Tarski's background. In: J. Hintikka (Ed.), *From Dedekind to Gödel. Essays on the Development of the Foundations of Mathematics*, Kluwer Academic Publishers, Dordrecht, pp. 331–341. Reprinted in: Woleński (1999), 126–133.

Woleński, J. (1996). Reism in the Brentanian tradition. In: L. Albertazzi et al. (Eds.), *The School of Franz Brentano*, Kluwer Academic Publishers, Dordrecht, pp. 357–375. Reprinted in: Woleński (1999), 179–190.

Woleński, J. (1997). *Szkoła Lwowsko-Warszawska w polemikach*, Wydawnictwo Naukowe Scholar, Warszawa.

Woleński, J. (1999). *Essays in the History of Logic and Logical Philosophy*, Jagiellonian University Press, Kraków.

Woleński, J. (2002). From intentionality to formal semantics (From Twardowski to Tarski), *Erkenntnis* 56, 9–27.

Woleński, J. (2003). Polish attempts to modernize Thomism by logic (Bocheński and Salamucha), *Studies in East European Thought* 55, 299–313. Reprinted in: J. Woleński *Historico-Philosophical Essays*, vol. 1, Copernicus Center Press, Kraków 2012, pp. 51–66.

Woleński, J. (2004). Analytic vs. synthetic and *a priori* vs. *a osteriori*. In: I. Niiniluoto, M. Sintonen and J. Woleński (Eds.), *Handbook of Epistemology*, Kluwer Academic Publishers, Dodrecht, pp. 781–839.

Woleński, J. (2009). Logic and the foundations of mathematics in Lwów (1900–1939). In: *Lwów Mathematical School in the Period 1915-1945 As Seen Today*, Institute of Mathematics, Polish Academy of Sciences, Warszawa, Banach Center Publications 87, 27–44.

Woleński, J., Simons, P. (1989). De Veritate: Austro-Polish Contributions to the theory of truth from Brentano to Tarski. In: K. Szaniawski (Ed.), *The Vienna Circle and the Lvov-Warsaw School*, Kluwer Academic Publishers, Dordrecht, pp. 391–442.

Zaremba, S. (1911). Pogląd na te kierunki w badaniach matematycznych, które mają znaczenie teoretyczno-poznawcze, *Wiadomości Matematyczne* 15, 217–23.

Zaremba, S. (1912). *Arytmetyka teoretyczna*, Polska Akademia Umiejętności, Kraków.

Zaremba, S. (1923). O stosunku wzajemnym fizyki i matematyki, in: *Poradnik dla samouków*, vol. III: *Matematyka. Uzupełnienia do tomu pierwszego*, A. Hefler and St. Michalski, Warszawa, pp. 131–167.

Zaremba, S. (1926). *La logique des mathématiques*, Mémorial des Sciences Mathématiques, XV, Gauthier-Villars, Paris, pp. 1–52.

Zawirski, Z. (1914). *O modalności sądów*, Nakładem Polskiego Towarzystwa Filozoficznego, Lwów.

Zaremba, S. (1938). Uwagi o metodzie w matematyce i fizyce, *Przegląd Filozoficzny* 41, 31–36.

Żyliński, E. (1921–1922). O przedmiocie i metodach matematyki współczesnej, *Ruch Filozoficzny* 6, 71a–71b.

Żyliński, E., *et al.* (1924). Memorial profesorów: E. Żylińskiego, H. Steinhausa, St. Ruziewicza i S. Banacha w sprawie studjum matematycznego na Wydziale Filozoficznym Uniwersytetu Jana Kazimierza we Lwowie adresowany do Departamentu Nauki i Szkół Wyższych Ministerstwa Wyznań Religijnych i Oświecenia Publicznego, Lwów, 14 kwietnia 1924 r., Archiwum Instytutu Matematycznego Polskiej Akademii Nauk w Sopocie.

Żyliński, E. (1925). Some remarks concerning the theory of deduction, *Fundamenta Mathematicae* 7, 203–209.

Żyliński, E. (1927). O przedstawialności funkcyj prawdziwościowych jednych przez drugie, *Przegląd Filozoficzny* 30, z. IV, 290.

Żyliński, E. (1928). Z zagadnień matematyki. II. O podstawach matematyki, *Kosmos*, Seria B, 53, 42–53.

Source Note

On the philosophical meaning of reverse mathematics, in: *Philosophy of Mathematics, Proceedings of the 15th International Wittgenstein-Symposium*, Hrsg. J. Czermak, Verlag Hölder-Pichler-Tempsky, Wien 1993, pp. 173–184.

On the distinction proof–truth in mathematics, in: P. Gärdenfors *et al.* (Eds.), *In the Scope of Logic, Methodology and Philosophy of Science*, Kluwer Academic Publishers, Dordrecht–Boston–London 2002, pp. 287–303.

Some historical, philosophical and methodological remarks on proof in mathematics, in: D. Probst and P. Schuster (Eds.), *Concepts of Proof in Mathematics, Philosophy, and Computer Science*, Ontos Mathematical Logic, Walter de Gruyter, Berlin 2016, pp. 251–268.

The status of Church's thesis, in: *Church's Thesis After 70 Years*, A. Olszewski, J. Woleński, R. Janusz (Eds.), Ontos Verlag, Frankfurt/Main 2006, pp. 310–330 (co-author: J. Woleński)

Between theology and mathematics. Nicholas of Cusa's philosophy of mathematics, *Studies in Logic, Grammar and Rhetoric* 44 (57) (2016), 97–110.

Phenomenological ideas in the philosophy of mathematics. From Husserl to Gödel, *Studia Semiotyczne* 32, nr 2, 2018, 29–47 (co-author: Th. Bedürftig).

Tarski and his Polish predecessors on truth, in: D. Patterson (Ed.), *New Essays on Tarski and Philosophy*, Oxford University Press, Oxford 2008, pp. 21–43 (co-author: J. Woleński).

Benedykt Bornstein's philosophy of logic and mathematics, *Axiomathes* 24 (2014), 549—558.

Philosophy of logic and mathematics in the Warsaw School of Mathematical Logic, *Studies in Logic, Grammar and Rhetoric* 27 (40) (2012), 145–155.

Philosophy of mathematics and logic in Cracow between the wars, in: K. Mulligan, K. Kijania-Placek, T. Placek (Eds.), *The History and Philosophy of Polish Logic. Essays in Honour of Jan Woleński*, Palgrave Macmillan, New York/London 2014, pp. 278–294.

Philosophy of logic and mathematics in the Lwów School of Mathematics, *Mathematical Bulletin of Taras Shevchenko Scientific Society* 10 (2013), 17–24.

Cracow Circle and its philosophy of logic and mathematics, *Axiomathes* 25 (2015), 359–376.

Index

Achtner W. 181
Ackermann W. 13, 27, 47, 176, 181, 188
Agazzi E. 182
Aigner M. 42, 181
Ajdukiewicz K. 134, 177, 181, 182
Albertazzi L. 202
Albertson D. 181
Albertus Magnus 69
Appel 42, 43
Apt K.R. 16, 181
Aristotle 23, 38, 76, 109–112, 118, 168, 178, 192
Arzelà C. 19
Aschbacher M. 42, 181
Ascoli G. 19
Avigad J. 40, 42, 181

Banach S. 6, 18, 103, 107, 159, 160, 162, 181, 183, 202, 203
Banachiewicz T. 96
Barnett D.I. 191
Barwise J. 50, 181, 195, 198
Bassler G.B. 43, 181
Becker O. 5, 6, 86–88, 181
Benacerraf P. 93, 187, 191
Bendiek J. 166, 167, 182
Bergmans L. 184
Bernays P. 13–16, 182, 188, 199
Berry G.D. 39
Betti A. 108, 182
Blackwell K. 193, 194
Bocheński J.M. 6, 102, 151, 163–167, 170, 172, 177–180, 182, 202
Bolzano B. 19, 39, 81, 82, 108, 164
Borkowski L. 192
Born M. 149
Bornstein B. 6, 9, 127–136, 182, 183, 191, 198, 204
Bourbaki N. 49, 183
Bradwardine Th. 70

Breger H. 192
Breidert W. 183
Brentano F. 81–83, 108, 109, 113–115, 196, 202
Brodie H.C. 184
Browder F.E. 21, 183, 188
Brown D.K. 18, 183
Burali-Forte C. 39

CadwalladerOlsker T. 41, 49, 183
Campanus of Navara 70
Cantor G. 39. 81–85, 89, 90, 96, 130, 141, 152, 183, 187
Carnap R. 29, 46, 62–64, 89, 92, 124, 183, 192
Cauchy A.L. 18, 19, 39, 50, 82, 197
Chihara Ch. 93, 123
Church A. 5, 9, 37, 48–50, 53–70, 182, 183, 186, 189–191, 193–195, 197, 204
Chwistek L. 6, 98, 102, 103, 124, 137, 138, 145, 148, 151–157, 159, 162, 169, 184, 194
Clark J.T. 166, 167, 184
Coleman A.P. 184
Coniglione F. 201
Corcoran J. 199
Counet J.M. 184
Couturat L. 112, 184
Czermak J. 185, 204
Czeżowski T. 6, 102, 108, 112, 119, 184

Davis M. 46, 50, 183, 184, 186, 187, 196
De Long H. 55, 184
De Villiers M.D. 41, 185
Dedekind R. 11, 12, 39, 82, 83, 85, 201
Denjoy A. 157, 191
Detlefsen M. 11–13, 184, 185, 189
Drake F.F. 17, 185
Drewnowski J.F. 6, 151, 163–167, 170, 172–177, 185

Edwards P. 182, 193
Ehrenfeucht A. 195
Einstein A. 149
Erdös P. 42
Euclid 23, 38–41, 45, 70, 84, 129, 147, 149, 155, 156, 171
Eugenius IV 69, 70
Euler L. 135

Federici Vescovini G. 185
Feferman A.B. 15, 107, 123, 124, 140, 185, 187
Feferman S. 15, 29, 107, 123, 124, 140, 185, 187
Field H. 115, 185
Fløstad G. 196
Folina J. 54, 57, 186
Føllesdal D. 88, 89, 91, 186
Fourier J. 39
Fréchet M. 97
Frege G. 11, 12, 24, 39, 44, 45, 81, 85, 86, 93, 186, 187, 194, 199
Friedman H. 16–21, 186, 197
Frost-Arnold G. 124, 186

Gabriel G. 186
Gałecki Ł. 7
Gandy R. 53, 58, 186
Gildner C. 156
Givant S.R. 199
Gödel K. 5, 6, 9, 13–15, 18, 20, 26–30, 33, 36, 45–48, 59, 60, 81, 88–93, 105, 123, 140, 142, 180, 184–187, 191, 194, 200, 201, 204
Goodman N. 182
Grassmann H. 81, 135
Grattan-Guinness I. 81, 187
Grelling K. 39
Grosholz E. 192
Grover D. 115, 187
Grzegorczyk A. 193

Haaparanta L. 199
Hahn H. 18, 183
Haken W. 42, 43

Haller R. 199
Hardy G.H. 42
Harrington L. 14, 20, 92, 195, 197
Hartimo M. 88, 187
Hausdorff F. 96
Heijenoort J. Van 187, 188, 201
Heller M. 181
Hermes H. 186
Hersh R. 42, 187
Herzberg J. 156
Hetper W. 156
Heyting A. 189, 193
Hilbert D. 5, 11–17, 19–21, 23–27, 29, 30, 36, 39, 44–47, 50, 81, 86, 91, 93, 117, 181–183, 185, 187, 188, 191, 196–198, 201
Hintikka J. 186, 201
Hiż H. 73, 166, 168, 189
Hölder O. 87, 185, 199, 204
Horwich P. 115, 189
Huntington E.V. 134
Husserl E. 5, 5, 9, 81–93, 153, 186, 187, 189, 199, 204

Isaacson D. 36, 189

Jadacki J.J. 192, 197
Jadczak R. 200
Janiszewski Z. 96, 97, 99–101, 139, 149, 159, 189
Janusz R. 195, 204
Jaśkowski S. 102, 146, 193
John II Casimir Vasa 95
Jordan Z. 102

Kaczmarz S. 159
Kahle R. 42, 189
Kalmár L. 54, 64, 189
Kambartel F. 186
Kant I. 11, 12, 67, 85, 90, 127, 149, 155, 177, 185
Kauferstein C. 85, 189
Kaufmann F. 88, 189
Kaye R. 32, 189
Kijania-Placek K. 204

Kino A. 191
Kirby L. 14, 20, 92, 189
Kleene S.C. 54, 57–60, 103, 186, 189, 190
Klibansky R. 195
Knobloch E. 190
Koj L. 168, 169, 190
Kokoszyńska M. 119, 190
König D. 17, 18
Kopelman K. 156
Kotarbińska J. 190
Kotarbiński T. 6, 104, 107, 110–121, 123, 124, 128, 139, 140, 163–166, 189, 190, 196, 201
Kotlarski H. 34, 35, 190
Krajewski S. 33, 58, 191
Krazer A. 188
Kreisel G. 13, 15, 54, 56, 191
Kronecker L. 81
Kuratowski K. 103, 105, 141, 148, 167, 191
Kuzawa S.M.G. 189, 191

Lachlan A. 189
Lebesgue H. 97
Leibniz G.W. 39, 44, 88, 134, 135
Leśniewski S. 97, 98, 101–104, 107, 108, 110, 113, 114, 117, 119, 121, 124, 126, 128, 130, 132, 133, 136–139, 152, 157, 163, 164, 182, 191, 198
Lindenbaum A. 101, 103, 105
Lindström S. 198
Linsky B. 194
Llull Ramon 69
Lusin N. 97, 157, 191
Łukasiewicz J. 6, 97, 98, 101, 102, 104, 105, 107, 108, 110–113, 118, 119, 121, 122, 138, 139, 148, 150, 163, 164, 166, 167, 172, 191, 192

Mach E. 154
Maddy P. 93, 192
Maligranda L. 192
Mancosu P. 42, 86, 124, 192
Marczewski E. 99, 193
Marek W. 16, 181, 191, 196
Mazur S. 103, 159, 193

McCarty C. 54, 193
McKenzie R.N. 199
Melamid A. 156
Mendelson E. 20, 50, 56–58, 62, 64, 193
Menne A, 182
Michalski K. 164
Michalski St. 202
Mleczko P. 7
Mostowski A. 6, 96, 101, 103, 105, 123, 124, 139–143, 191, 193, 194, 195, 198, 199
Mulligan K. 204
Murawski R. 7, 16, 28, 32, 33, 44, 45, 50, 53, 54, 67, 81, 92, 96, 125, 126, 139, 149, 151, 152, 157, 161, 169, 181, 182, 193–195
Mycielski J. 124, 195
Myhill J. 191

Neurath O. 199
Nicholas V 70
Nicolai de Cusa (also known as: Nikolas of Cusa, Cusanus) 5, 9, 69, 71, 72, 74–79, 181, 183–185, 195, 204
Niebergall K.-G. 32, 195
Nieznański E. 167, 195
Niiniluoto I. 202
Nikliborc W. 159
Nikolas Krebs (also known as: Nikolas Kryffs) 69

Ockham W. 196
Olszewski A. 186, 193, 204
Orlicz W. 159

Paris J. 14, 20, 92, 189, 195
Parsons Ch. 93, 195
Pascal B. 154
Pasquerella L. 114, 196
Patterson D. 182, 204
Pauli W. 105
Peano G. 14, 16–20, 31–36, 50, 61, 65, 85, 189, 194, 195, 197
Pearce D. 190

Peckhaus V. 192
Peirce Ch.S. 135
Pepis J. 103, 156
Placek T. 204
Plato 23, 38, 73, 115, 153
Pogonowski J. 194
Poincaré H. 16, 97, 127, 129, 156,
Poli R. 201
Policki K. 167
Post E. 14, 55, 150, 196
Poznański E. 117
Prawitz D. 12, 196
Presburger M. 101, 103, 105
Probst D. 204
Pseudo-Dionysius 69
Putnam H. 187, 191
Puzyna J. 96, 99
Pythagoreans 39, 73

Quine W.V.O. 65, 116, 124, 192, 196,

Ramsey F.P. 196
Rasiowa H. 193
Ratajczyk Z. 20, 34, 35, 190
Resnik M.D. 12, 13, 196
Riemann B. 18
Rivetti Barbò F. 167
Robinson R.M. 199
Rogers H. 54, 57-59, 196
Rohn K. 87
Rojszczak A. 114, 196
Rosser J.B. 186
Rota G.-C. 42, 196
Rowe D.E. 23, 39, 83, 196
Russell B. 39, 44, 45, 89, 102, 103, 110, 121, 124, 151, 166, 167, 179, 187, 194, 201
Ryckman Th. 86, 192
Ryle G. 63, 64, 196
Rynin D. 113, 190

Salamucha J. 6, 102, 151, 163-172, 195-197, 202,
Schauder J. 159

Schilpp P.A. 183, 187
Schlick M. 149
Scholz H. 118, 179, 197
Schulz D.-K. 53, 62, 197
Schürmann A. 190
Schuster P. 204
Schütte K. 15
Segerberg K. 198
Shapiro S. 55, 57, 65, 67, 197
Sierpiński W. 96-99, 101, 103-105, 107, 113, 137, 142, 149, 157, 193, 197
Sigismund 70
Sikorski R. 193
Simons P. 108, 202
Simpson S.G. 15, 16, 18-20, 183, 186, 197, 198
Sinaceur H.B. 126, 198, 199
Singer H.-G. 195
Sintonen M. 202
Skarżyński J. 156
Skolem Th. 16, 20, 27, 47
Skolimowski H. 107, 198
Sleszyński J. 6, 145, 146, 150, 198
Smith R.L. 18, 19, 186,
Smoryński C. 11-13, 16, 198
Smullyan R. 59, 198
Sobociński B. 6, 164, 165
Srebrny M. 189
Srzednicki J. 195
Stachowiak M. 7
Steiger R. 195
Steinberżanka D. 190
Steiner M. 42, 93, 198
Steinhaus H. 6, 159-161, 198, 203
Stoltenberg-Hansen V. 198
Stumpf C. 81
Suppes P. 123, 150, 198
Surma S.J. 191
Suszko R. 141, 198
Sylvan R. 157, 183, 198
Szaniawski K. 182, 202
Śleziński K. 129, 132, 198

Świętorzecka K. 197
Świrydowicz K. 161, 194

Tait W.W. 12, 16, 198
Takeuti G. 15, 198
Tarski A. 6, 9, 28, 29, 32, 33, 36, 45–47, 50, 56, 64, 67, 101–105, 107, 108, 110–114, 116–126, 139–141, 159, 160, 167, 179, 181, 182, 185, 186, 190, 192–195, 198–202, 204
Tarski J. 120
Tatarkiewicz W. 163
Tega W. 185
Thomas Aquinas 165, 167, 179, 196
Tieszen R. 93, 199
Toscanelli Paolo dal Pozzo 69
Turing A. 37, 48, 49, 53, 58, 186
Twardowski K. 6, 96–98, 107–112, 115, 117–119, 122, 125–128, 138, 146, 159, 160, 172, 199, 200, 202,
Tymoczko Th. 43, 200

Ulam S. 103, 159

Venn J. 135
Veraart A. 186
Vesley R.E. 191
Vuissoz F. 108, 200

Waltuch K. 157
Wandycz D. 189
Wang Hao 26–29, 46, 88, 89, 200
Webb J. 55, 59, 200
Wechsung G. 194
Weierstrass C. 19, 39, 50, 56
Weingartner P. 185
Weiss B. 190
Weyl H. 5, 6, 16, 86–88, 105, 192, 200, 201
Whitehead A.N. 102, 103, 121, 151, 167, 201,
Wiedijk F. 48, 201
Wilkosz W. 6, 145, 150, 151, 201
Wolak Z. 164, 167, 190, 201
Woleński J. 7, 9, 28, 50, 53, 65, 96, 107, 108, 110, 113–116, 119, 120, 126, 127, 129, 136, 139, 148, 157, 159, 164, 190, 195, 200–202, 204
Woodin W. Hugh 181
Wundheiler A. 117

Zarach A. 189, 191
Zaremba S. 6, 99, 100, 145, 147, 148, 157, 192, 202
Zaremba S.K. 198
Zawirski Z. 6, 108, 112, 115, 145, 149, 150, 160, 202
Zermelo E. 20, 30, 81, 96, 132, 197
Ziegler G.M. 42, 181
Żorawski K. 99
Żyliński E. 6, 159, 161, 162, 192, 203

Polish Contemporary Philosophy and Philosophical Humanities

Edited by Jan Hartman

Vol. 1 Roman Murawski: Logos and Máthēma. Studies in the Philosophy of Mathematics and History of Logic. 2011.
Vol. 2 Cezary Józef Olbromski: The Notion of *lebendige Gegenwart* as Compliance with the Temporality of the "Now". The Late Husserl's Phenomenology of Time. 2011.
Vol. 3 Jan Woleński: Essays on Logic and its Applications in Philosophy. 2011.
Vol. 4 Władysław Stróżewski: Existence, Sense and Values. Essays in Metaphysics and Phenomenology. Edited by Sebastian Kołodziejczyk. 2013.
Vol. 5 Jan Hartman: Knowledge, Being and the Human. Some of the Major Issues in Philosophy. Translated by Ben Koschalka. 2013.
Vol. 6 Roman Ingarden: Controversy over the Existence of the World. Volume I. Translated and annotated by Arthur Szylewicz. 2013.
Vol. 7 Jan Hartman: Philosophical Heuristics. Translated by Ben Koschalka. 2015.
Vol. 8 Roman Ingarden: Controversy over the Existence of the World. Volume II. Translated and annotated by Arthur Szylewicz. 2016.
Vol. 9 Tomasz Kubalica: Unmöglichkeit der Erkenntnistheorie. Leonard Nelsons Kritik an der Erkenntnistheorie unter besonderer Berücksichtigung des Neukantianismus. 2017.
Vol. 10 Renata Ziemińska: The History of Skepticism. In Search of Consistency. 2017.
Vol. 11 Jan Woleński: Logic and Its Philosophy. 2018.
Vol. 12 Wielslaw Gumula: On Property and Ownership Relations. 2018.
Vol. 13 Andrzej Zaporowski: Action, Belief, and Community. 2018.
Vol. 14 Andrzej Bator/Zbigniew Pulka (eds.): A Post-Analytical Approach to Philosophy and Theory of Law. 2019.
Vol. 15 Krzysztof Śleziński: Towards Scientific Metaphysics. Volume 1: In the Circle of the Scientific Metaphysics of Zygmunt Zawirski. Development and Comments on Zawirski's Concepts and their Philosophical Context. 2019.

Vol. 16 Krzysztof Śleziński: Towards Scientific Metaphysics. Volume 2: Benedykt Bornstein's Geometrical Logic and Modern Philosophy. A Critical Study. 2019.

Vol. 17 Jan Felicjan Terelak: Eustress and Distress: Reactivation. 2019.

Vol. 18 Roman Murawski: Lógos and Máthēma 2. Studies in the Philosophy of Logic and Mathematics. 2020.

www.peterlang.com